上海空中行走地图
Shanghai Skywalkers

著 王桢栋 / 冯宏 / 尹明 / 营光学社
By Wang Zhendong / Feng Hong
Yin Ming / Highlight Studio

译 营光学社 /(美)丹尼尔·萨法里克
Translated by Highlight Studio / Daniel Safarik

上海文化出版社
SHANGHAI CULTURE PUBLISHING HOUSE

浦江之晨
First Light on the Huangpu River

引言

每一个对城市的过去、现在和未来感兴趣的人,都应将上海作为首选必游之地。在全面引领大规模现代化进程的同时,上海亦是中国最具特色的大城市之一,在闪耀的摩天大楼和辉煌的购物中心的荫蔽下,狭窄的石库门弄堂也能泰然处之。作为通往中国和东方的真正门户,即便在经济高度全球化的今天,上海依旧保有"古老的灵魂"和深厚的历史,以及具有中国传统、西方艺术风格和装饰艺术派的建筑。

上海的活力是无与伦比的。在1993年第一条地铁开通后,轨道交通系统仅仅经过29年的建设便已跻身世界前列:这个拥有20条线路、509座车站、全长801公里的庞然大物,每天为1000多万人提供服务。几乎一夜之间,陆家嘴中央商务区从1990年代的低层轻工业区发展成为如今的世界级摩天大楼群,其中包括世界第三高楼上海中心大厦,以及数十座超过200米的高层建筑。

世界高层建筑与都市人居学会曾在2012年和2014年,极短的间隔内,两次选择上海作为全球大会举办地。因为在两年中,这座城市乃至整个中国发生了如此多的变化,有全新的议程可供讨论,还有令人振奋且富有创意的新建高层建筑可供参观。上海已成为中国改革开放进程中前所未有的建筑设计浪潮发源地。为借助上海这座国际大都市对国内外熟练劳动力的巨大吸引力,以及与广阔内陆地区良好的连通性,跨国企业纷纷在这里设立办事处。随着中国扩大其建筑设计和施工的全球专业市场,上海必将成为输出人才和资源的要地。

值得一提的是,学会在上海举办国际会议的同时,将亚洲总部落户同济大学。这个决定并非巧合,专业扎实、学术严谨以及包括本书作者在内的教授们友好热忱的合作态度,使之成为学会又一个毋庸置疑的正确选择。上海已成为学会乃至整个高层建筑行业的大本营。我们很高兴与世界一起分享本书所呈现的独特视角,它深刻诠释了上海这座城市的活力所在。请享受你们的上海"空中漫步"!

安东尼·伍德
世界高层建筑与都市人居学会主席

FOREWORD

In many ways Shanghai is at the leading edge of China's massive modernization project, it also remains one of the nation's most charactertical large cities, with its narrow shikumen alleys coexisting in the shadow of sparkling skyscrapers and glittering shopping centers. Long the true front door to China and the East, it is a city with an "old soul" and a deep history, steeped in traditional Chinese, Western classical and Art-Deco architecture, and it has remained so even amidst today's highly globalized economy.

The dynamism of Shanghai is unrivaled. Its metro system launched its first line in 1993, and in the space of 29 years has grown into the world's largest: a 20-line, 509-station, 801-kilometer behemoth serving more than 10 million people each day. The Lujiazui CBD sprung up practically overnight, from a low-rise light-industrial district in the 1990s to a world-class skyscraper cluster today, including Shanghai Tower, the world's third-tallest building, and dozens of other towers over 200 meters in height.

The Council on Tall Buildings and Urban Habitat chose Shanghai as the location for its International Conference in both 2012 and 2014. For within the space of those two years, so much had changed in the city and in China at large, that there was an entirely new agenda to discuss, and a new cohort of exciting and innovative tall buildings to tour. Shanghai also had become a launchpad for the unprecedented wave of architectural design work afforded by China's Reform and Opening-up process. International firms established offices here to draw on the cosmopolitan city's great attractiveness to skilled labor, both from abroad and domestically, and its excellent connectivity to the nation's vast interior.

It is not a coincidence that, in addition to selecting the city as a conference site, the Council chose Shanghai, and Tongji University as the location for its Asia Headquarters office. The confluence of architectural knowledge, academic rigor, and the welcoming, collaborative attitude of its administrators and professors – including the authors of this book – made this an obvious choice. For all these reasons, Shanghai is a home base for the Council and for the entire tall building industry. We are pleased to share this unique perspective of Shanghai with the world, as it is very much reflective of all these dynamics converging. Enjoy your Shanghai "skywalk"!

President, Council on Tall Buildings and Urban Habitat

序 行走在城市的上空

在神话中，人们幻想着乘坐飞毯在城市中自由穿行。1980年有过飞毯似的个人飞行器设想，在城市中飞速穿行。1982年的电影《银翼杀手》里的警车就在城市上空飞行，许多未来理想城市的交通想象就是飞行汽车。《上海空中行走地图》（以下简称《行走地图》）让你体验一下在城市中飞行的乐趣，从特殊的角度来阅读上海这座摩天大都市。

真要想在布满高层建筑的垂直城市中行走，一个人大概需要有近千米的巨人身高，飞行才是唯一可行的方式。实际上，书中的行走宛若飞行，不受道路的限制，可以随心所欲地从高空俯瞰，多视角并立体化地观赏城市这一特殊的大地艺术，还能不断地发现"新大陆"。以高空视野穿行在上海中心城区的空中公共空间，必然带给人们视觉的冲击和惊喜。《行走地图》的飞行路线设定的是探访上海的高层建筑地标，以公共开放空间为主线，"飞行者"如大鹏鸟驻足于建筑上，极目远望，阅尽人间风光。《行走地图》谈历史，谈建筑，谈城市交通，谈城市生活和它的烟火气，谈城市更新；有赞赏，有解说，有感想，也有评论。

《行走地图》的特点是精美的空中摄影，许多视角是日常生活中不可遇见的。此外，与精致的图片相匹配，书中的文字也是作者经过推敲尽可能做到词语达意。有一位学者说过：影像往往比实际事物更为真实，因为，摄影是通过"照相机的眼睛"和摄影的理念教会人们一种新的观察世界的规则。摄影既反映又创造现实，阅读摄影就是去理解摄影对象及其周围环境的关联、摄影对象的各个要素与社会文化的渊源；去理解图像的内涵，揭示掩藏在摄影话语背后存在的现实以及摄影所创造的现实。摄影不仅是纪实和表现，建筑摄影通过新的视觉方式创造世界。

许多城市和地区都有空中摄影图集，我也喜欢收集这类图集。北京有一本《天下北京》，上海有四五种空中看上海的图集，一般都是在空中巡游，而《行走地图》是引领读者深入公共空间并解读城市空间。

首先是回溯城市的历史，《行走地图》开篇就把人们的视线带入外滩。外滩是上海的地标和历史的见证，之后就是老城厢——上海的城市之根，一路尽情欣赏之余，飞到南外滩，俯瞰黄浦江，对视浦东陆家嘴的中央商务区。从南外滩飞往苏州河，进入苏河湾，去看看大悦城的摩天轮。接着飞往上海的另一个地标——南京西路的新世界丽笙大酒店，从大酒店的顶端一睹中华第一街南京东路的纷繁多彩。接下来，从中央活动区又飞向另一个中央活动区——徐家汇，阅读城市的历史变迁。

在第二部分里，《行走地图》带着读者飞回到那座火箭般造型的明天广场。这里可以对城市空间的演变作一番讨论了。驻足上海的中心——人民广场的上空,从人们通常不可能到达的视角观赏上海博物馆、上海大剧院,以及上海历史博物馆。飞到环球港,便可以俯视 1950 年代的地标——曹杨新村。接下来从长宁龙之梦,俯瞰绿意朦胧的中山公园。现在来到芮欧百货,静安寺和静安公园尽收眼底。接着,来到 1988 年建成的新锦江大酒店,上海申字形高架道路网一览无余。改革开放后,为了促进城市发展,上海加速推进城市快速道路建设。在新锦江大酒店顶层的旋转餐厅眺目而望,直观地感受到高架道路建设对城市空间的影响。打浦桥地区是改革开放和旧区改造的早期先例,来到打浦桥地区,从斯格威铂尔曼酒店感受城市面貌和形态的持续变化,并从另一个角度回望黄浦江。

《行走地图》的第三部分是着重解读垂直生长的城市新面貌,以东方明珠电视塔、陆家嘴中央商务区、苏州河治理、北外滩滨江公共空间、后世博园区的当代艺术博物馆,以及虹桥商务区为例,向读者解说上海的城市空间变化,展示功能转型时期的城市空间演变。

这本书的作者是一群有着建筑学专业背景的师生,他们满怀理想,希望以他们的所学所知,为公众提供阅读城市的新视角。他们以精心挑选的"新沪上 18 景"的公共开放空间为观察点,在通过航拍摄影向读者呈现独特空中视野的同时,叙述城市的发展脉络,从不同的角度为读者提供阅读上海和感知上海的新方式。

PREFACE Walk Over the City

In many mythology stories, people have imagined traveling freely through the city on a magic flying carpet. In 1980, there was an idea of flying carpet-like personal aircraft soaring across the sky; in 1982, police cars appeared shuttling aerially in the movie *Blade Runner*. Flying automobiles have become the form of future transportation in all kinds of fantasies. *Shanghai Skywalkers* allows you to experience the fun of "flying" and to "read" Shanghai, from a special perspective.

Unless one is a 1,000-meter-tall giant, flying is the only way to navigate a city full of skyscrapers. In fact, walking on the "*Shanghai Skywalkers*" is just like flying without obstacles. The walkers will see the city as a great piece of "Land art" in three dimensions from multiple perspectives. During the leisurely wandering in the sky, "new continents" might be discovered in the populated city. Walking through the aerial public space in the center of Shanghai, the broad views from a high altitude definitely will bring people surprises and new visual impact. The "flight routes" in *Shanghai Skywalkers* cover many skyscraper landmarks in Shanghai. Standing on the public space of high-rise buildings, people can look much further like a bird overlooks the city vastly. It seems that nothing is left out of the view and the whole world is in your eyes. Here, we talk about the history of this city, historical buildings, hustle and bustle, traffic, urban life, and its renewal, with appreciations, explanations, impressions, and comments.

Shanghai Skywalkers contains many precious photographic works, some of which are taken from a rare perspective and hardly can be reached in daily life. A scholar once said: "Photography is sometimes more authentic than reality, through the camera, photography teaches people a new method to observe the world." Photography is reflecting and creating a reality at the same time; therefore, "reading" photographs is to understand how to relate the objects in camera lens with the surrounding environment and culture; more importantly, it is to understand the connotation of the images, to reveal the reality hidden behind photographs and the reality created by photography. From this perspective, photography is not only documentary or performance; architectural photography creates reality through new visual representations.

Aerial photography atlases exist for many cities and regions, and I like to collect them as well. Usually, they are captivating from an aerial perspective, but *Shanghai Skywalkers* goes much deeper into the city public space and much better explains and interprets the significance of city public space.

First of all, *Shanghai Skywalkers* reviews the history by viewing the Bund. The Bund is an important landmark of Shanghai. It has witnessed the history of Shanghai for hundreds of years. The next stop is the Old Town area. That is the root of Shanghai. After visiting these places, we fly to the South Bund to overlook the Huangpu River and look at the Lujiazui CBD in Pudong. We then proceed to the Suzhou Creek to visit the Ferris wheel in the Jing'an Joy City. Afterwards, we fly to the Radisson Blu Hotel Shanghai New World, another landmark of Shanghai, on West Nanjing Road. On the top roof of the hotel, we can overlook "the first street of China, East Nanjing Road." The following spot will be another central activity area in Shanghai, Xujiahui, where we can study the historical changes of the city.

Subsequently, back to the rocket-shaped Tomorrow Square *Shanghai Skywalkers* focuses on the modern evolution of urban space. The People's Square is the center of Shanghai, and over it we can overlook the Shanghai Museum, the Shanghai Grand Theatre, and the Shanghai History Museum from different perspectives that people wouldn't normally see. Inadvertently, we shift our sight to the Global Harbor, where we can look down at the Caoyang Community, a landmark of the 1950s. Later, we fly to Cloud Nine, enjoying the scenery of Zhongshan Park in the haze. Now we arrive at Réel Department Store, Jing'an Temple and Jing'an Park are in the views. At last, we land onto the Jin Jiang Tower Hotel which was built in 1988. The "申" (shen, the Chinese abbreviation of Shanghai)-shaped elevated road network is a complete entirety in our eyes. After the Reform and Opening-up, Shanghai accelerated the construction of urban highways to further promote urban development. From the revolving restaurant on the top floor of Jin Jiang Tower Hotel, we can directly feel the impact of the construction of the elevated road on the urban space. The Dapuqiao area is a precedent for the Reform and Opening-up and the reconstruction of old districts. Here, we can feel the changes of the city from the Pullman Shanghai Skyway Hotel, and look back at the Huangpu River from another angle.

The third part of *Shanghai Skywalkers* interprets the new appearance of the vertically growing city, taking in the Oriental Pearl Radio & TV Tower, Lujiazui CBD, the governance of Suzhou Creek, the North Bund public space, Power Station of Art in the Post-World Expo Park, and the Hongqiao CBD as examples; each of them shows the evolution of urban space in the period of transformation.

The authors of this book are a group of passionate and aspiring teachers and students majoring in architecture. Taking 18 key public open spaces as observation points, they use photographs to present readers a unique aerial view while to narrate the development of the city and provide a new perspective to read and perceive Shanghai.

Zheng Shiling

目录

引言···安东尼·伍德

序　行走在城市的上空·································郑时龄

第1章　水平扩张时期的城市风貌遗存 ············· 13
和平饭店：外滩建筑群································17
豫园商城：老城厢的演变······························25
外滩英迪格酒店：黄浦江跨江系统······················35
静安大悦城：铁路建设的起点··························43
新世界丽笙大酒店：南京路的商业发展··················51
美罗城：徐家汇的变迁································59

第2章　功能转型时期的城市空间演变 ············· 67
明天广场：人民广场的改造····························71
环球港双子塔：工人新村的诞生························79
长宁龙之梦：公共服务设施的蜕变······················87
芮欧百货：静安寺的前世今生··························95
新锦江大酒店：高架道路的建设······················· 105
斯格威铂尔曼酒店：打浦桥地区旧城更新················ 115

第3章　垂直生长时期的城市文明新貌 ············ 125
东方明珠和上海中心：陆家嘴的高度攀升················ 128
宝格丽酒店：苏州河水岸治理························· 141
外滩茂悦酒店：城市滨水空间························· 151
上海当代艺术博物馆：后世博园区的发展··············· 161
白玉兰广场：北外滩的新生··························· 169
虹桥交通枢纽：虹桥商务区的崛起····················· 179

后记　迈向人民的垂直城市··························· 188

CONTENTS

FOREWORD .... Antony Wood
PREFACE Walk Over the City .... Zheng Shiling

CHAPTER 1   The Urban Style and Features during the Period of Horizontal Expansion 14

Peace Hotel: The Buildings on the Bund 17
Yuyuan Bazaar: The Evolution of the Old Town Area 25
Hotel Indigo on the Bund: The Cross-river System of Huangpu River 35
Jing'an Joy City: The Starting Point of Railway Construction 43
Radisson Blu Hotel Shanghai New World: The Commercial Development of Nanjing Road 51
Metro City: Changes of Xujiahui 59

CHAPTER 2   The Evolution of Urban Space during the Transformation Period 68

Tomorrow Square: The Transformation of People's Square 71
Global Harbor: The Founding of New Villages 79
Cloud Nine: The Construction of Public Service Facilities 87
Réel Department Store: A Memory of Jing'an Temple Area 95
Jin Jiang Tower Hotel: The Construction of Elevated Roads 105
Pullman Shanghai Skyway Hotel: The Urban Regeneration of Dapuqiao Area 115

CHAPTER 3   The New Look of Urban Civilization in the Period of Vertical Growth 126

Oriental Pearl Radio & TV Tower and Shanghai Tower: The Rising of Lujiazui CBD 128
Bulgari Hotel Shanghai: The Comprehensive Environmental Treatment of the Suzhou Creek 141
Hyatt on the Bund: Open Space of the Huangpu Waterfont 151
Power Station of Art: The Development of the Post-World Expo Park 161
Sinar Mas Plaza: The Rebirth of the North Bund 169
Hongqiao Transportation Hub: The rise of Hongqiao CBD 179

POSTSCRIPT Towards a Vertical City for the People 189

第 1 章
水平扩张时期的城市风貌遗存

上海凭借其优越的地理位置，在开埠后一跃成为近代中国重要的通商口岸，工商业飞速发展。从外滩、老城厢、十六铺码头、老北站、南京路和徐家汇的上空，可以沿着城市发展的时间线索感受近代上海空间风貌遗存。

1843 年上海开埠通商，英、美、法等国相继在老城厢外设立租界，分割地块、开辟道路、兴建各类建筑，促使传统的江南水乡开启了近代城市化进程。起初，租界仅以洋人居住地的形式存在，直至 1854 年工部局成立，允许华洋杂居，并慢慢发展成为拥有独立司法和行政权力的地区。

随后的几十年间，租界以"越界筑路"的方式不断扩张，在 1895 年前后达到高潮。1914 年，租界总面积已扩大至最初的 57 倍之多。扩张区域原先多为郊野荒地，其实质上推动了城市规模的快速扩大，客观上促进了城市交通发展。与此同时，英美租界的北部和西部成为工业发展的重点区域，今苏州河沿岸、杨树浦和北外滩等地的工业遗存正是这一时期的建设成果。租界的建设浪潮也影响了国人，乡绅们纷纷发起地方自治运动，在今老城厢等华界区域效仿租界开展城市建设活动。

就此上海近现代城市格局已具雏形，中外交融催生出独特的海派文化与城市风貌。由于上海在第一次世界大战中特殊的社会环境，在外国资本和早期民族商业资本的刺激下，租界的繁荣程度远高于其周围地区，南京路等地成为闻名遐迩的商业中心。这一时期也是上海乃至中国高层建筑建设的开端。在 1910 年代就已建成麦边大楼（亚细亚大楼）等高层建筑，大多在 6~8 层，其露台和屋顶餐厅是当时的城市"空中客厅"。在 1930 年代 10 层以上的高层建筑在上海出现以前，距离外滩不远的圣三一堂钟楼曾经是上海的制高点和最醒目的地标。1929 年建成 12 层的沙逊大厦（现和平饭店北楼），享有"远东第一楼"之称；1934 年南京路上建成 24 层的国际饭店，并保持了长达半个世纪的"上海第一高楼"记录；1935 年外白渡桥以北建成装饰艺术风格的百老汇大厦（现上海大厦）；1944 年建成外滩唯一具有中国传统建筑装饰的中国银行大楼。

高层建筑的高度，在不同历史阶段和不同生产力条件下，有着不同的标准。近代上海的高层建筑，不仅是上海摩天城市的风貌起源，也引领着中国高层建筑的发展。

CHAPTER 1
The Urban Style and Features during the Period of Horizontal Expansion

Relying on its superior geographical location, Shanghai became an important trading port in modern times of China. The booming business at the port drove rapid development of industry and commerce. Along the route from Shiliupu, Shanghaibei Railway Station, Nanjing Road, all the way to Xujiahui, the significant evidence reveals the trace of modern urban style in Shanghai in a bird's view over the Bund.

In 1843, Shanghai opened the trading port. Britain, the United States, France, and other countries successively established concessions outside the Old Town. At first, the concessions had beeen occupied by foreigners only until the establishment of the Shanghai Municipal Council in 1854. Each concession eventually had gained independent judicial and administrative powers.

In the following decades, the concessions continuously expanded their territories by building roads across the borders, and this activity had reached a peak around 1895. Till 1914, the total area of the concessions had been expanded to 57 times the original size. Since most of the expanded areas were originally outskirts and wastelands, the rapid expansion of the city's scope and the development of urban transportation had been essentially and objectively promoted. At the same time, the northern and western parts of the British and American concessions became major areas for industry. The industrial heritage in the areas along the Suzhou Creek, Yangshupu, and the North Bund are the results of construction during this period. The constructions in concessions had also affected the life of Chinese people. The local squires had launched a local autonomy movement to follow the example of concessions in urban construction activities in today's Old Town and other Chinese territories.

Thus, the modern urban structure of Shanghai began to form. The integration of Chinese and foreign influences gave the birth to Shanghai a unique culture and urban style. Not being affected by the World War I, the foreign capital and early national commercial capital stimulated the concessions in Shanghai into a concentrated commercial place; and the Nanjing Road and other places were formed as commercial centers with strong domestic and international reputations. This period was also the beginning of high-rise building constructions in Shanghai as well as in China. High-rise buildings, such as the McBain Building (the former Asiatic Petroleum Co.Building), built in the 1910s, mostly have an average of six to eight floors. The terraces and rooftop restaurants were the lounges with the views over the city. Before the

1930s, the tower of Holy Trinity Church was the highest and outstanding point of Shanghai. In 1929, the twelve-storey Sassoon House (now the North Building of the Peace Hotel) came to be known as "the highest Building in the Far East." In 1934, the Park Hotel was built on Nanjing Road, which had been the tallest building in Shanghai for half a century; In 1935, the Art Deco style Broadway Mansions was built on the north end of Waibaidu Bridge. In 1944, the Bank of China Building, the only building decorated with Chinese traditional pattern, was on the Bund.

As productive forces varied throughout the different stages, the standards of high-rise buildings have been keeping changing. These buildings of modern Shanghai were not only the origin of Shanghai's skyscraper city, but also have led the development of high-rise buildings in China.

9F
开放空间 Open Space

和平饭店北楼
The North Building of Peace Hotel

和平饭店：
外滩建筑群
Peace Hotel: The Buildings on the Bund

和平饭店的露台位于九层，这里既可纵览外滩的景色，亦可眺望陆家嘴建筑群，浦江两岸的繁华壮丽近在咫尺。

The terrace of the Peace Hotel is located on the 9th floor, with a wide view overlooking the buildings on the Bund and in Lujiazui.

建筑师	公和洋行
建成时间	1929 年
建筑功能	酒店
建筑高度	77 米
建筑层数	12 层
地址	南京东路 20 号
开放时间	12:00—22:30
公共交通	地铁 2、10 号线南京东路站；公交 37 路中山东一路北京东路站，公交 33、55 路中山东一路汉口路站
Architect	Palmer & Turner Group (P & T)
Built	1929
Building features	Hotel
Total building height	77m
Total floors	12
Address	20 Nanjing Rd. (E)
Recommended viewing time	12:00-22:30
Public Transportation	Metro Line 2/10 East Nanjing Rd. Station; Bus 37 Zhongshan Rd.(E-1)/Beijing Rd.(E) Station; Bus 33/55 Zhongshan Rd.(E-1)/Hankou Rd. Station

作为上海近现代城市空间的起点,外滩见证了上海近两百年的风云变幻。1843年上海开埠之初,外滩只是上海县城北面的一片滩涂。来沪外侨在外滩沿江建起简易平房,一字排开用作仓库、办公室和卧房。1843—1895年间,外滩建筑多为1~2层的券廊式砖木混合结构建筑。1895—1919年间,外滩建筑采用钢筋混凝土结构翻建,建筑高至5层以上,洋行、银行等办公建筑过半,为古典复兴和折衷主义建筑风格,内外装修讲究、设施增多。1908年建成的汇中饭店(现和平饭店南楼)拥有上海最早的电梯和屋顶花园。

20世纪20—30年代,外滩再次兴起翻建热潮。随着钢筋混凝土框架结构和钢架结构传入中国,新建建筑日渐增高,出现立面简洁的早期现代主义建筑。汇丰银行大楼(1923,现浦东发展银行)、海关大楼(1927)、沙逊大厦、百老汇大厦(1935)先后建成,代表了当时世界建筑的最新风尚,并与随后建成的中国银行大楼(1944)、交通银行大楼(1949,现上海市总工会)等形成外滩绝美天际线,拥有当时"远东华尔街"的金融地位。沙逊大厦也是上海近代建筑史上第一幢真正意义的现代主义建筑,其芝加哥学派的建筑外观和装饰艺术风格的室内设计,曾被称为"远东第一楼"。

如今站在和平饭店的露台上眺望,两旁是历经风霜的百年建筑群,对岸则是上海改革开放和社会主义现代化建设的成就——陆家嘴天际线。从清晨到子夜,外滩的十二时辰不只是海关大钟周而复始的轮回,更是沉淀在时光中的气象万千。作为见证历史、展示城市气质的"上海客厅",外滩将大气、现代与兼容并包的海派风范展现得淋漓尽致、流光溢彩。

As the starting point of Shanghai's modern urban space, the Bund has witnessed the vicissitudes of Shanghai over nearly 200 years. When Shanghai port was opened in 1843, the Bund was just a tidal flat in the north of the Old Town. Foreigners built simple bungalows along the riverside, used as warehouses, offices, and living spaces. In 1843-1895, most of the Bund buildings were masonry-timber structure in one or two floors of corridor with arches. From 1895 to 1919, the Bund buildings were renovated with reinforced concrete structures up higher than five floors of Classical Revivalism and Eclecticism style. Many of these buildings were used as offices for foreign firms and banks. Built in 1908, the Palace Hotel (now the South Building of Peace Hotel) was featured the earliest elevator and rooftop garden in Shanghai.

From 1920 to 1939, the reconstructions on the Bund had entered into a new phase. With the techniques of reinforced concrete frame structure and steel frame structure being introduced to China, new buildings were getting much taller than before, and early Modernism appeared. The HSBC Building

(1923, now SPDB Building), Customs House (1927), Sassoon House and Broadway Mansions (1935) built successively, reflected the modern trend of world architecture at that time. They were once the highest point on the Bund skyline. In 1944, the Bank of China Building was built, followed by the Bank of Communications Building completed in 1949. The completion of dozens of banks gave the Bund the name "Oriental Wall Street." The outstanding skyline of the "Buildings on the Bund" had formed. Sassoon House, influenced by Chicago School Art Deco style and reputed as "the highest building of the Far East," is the first modernism building in the history of Shanghai architecture.

Around the Peace Hotel, there stood elegant buildings of a century. On the opposite side of the Huangpu River, outstands the Lujiazui skyscrapers, the accomplishment of Pudong development. From early morning to midnight, the historic clock tower atop the Customs House strikes out the unique characteristics of Shanghai, a generous, compatible, and humble city.

外滩夜景
The Bund at night

和平饭店 | Peace Hotel

外滩信号塔与外滩建筑群
The Gutzlaff Signal Tower and the Buildings on the Bund

上海外滩建筑群

指位于上海浦西外滩黄浦江沿线的建筑，具有近百年历史，为古典复兴和折衷主义风格，以及早期现代主义风格，建筑色调基本统一，整体轮廓线协调。从上海开埠至今，外滩建筑群经历了3次大规模建造，保留至今的24栋优秀历史建筑在黄浦江边勾画出一道优美的天际线，有"万国建筑博览会"之誉。

The Buildings on the Bund

The Buildings on the Bund, aged almost 100 years, styled with Classical Revivalism, Eclecticism, and Early Modernism. They were shaped after three major constructions since the opening of Shanghai. 24 heritage buildings were preserved as a well known skyline of "the exotic building clusters" on the west bank of Huangpu River.

豫园商城：
老城厢的演变
Yuyuan Bazaar: The Evolution of the Old Town Area

豫园商城的海上梨园，隐于文昌路上和丰楼的四层。在其室外平台不仅可以俯瞰豫园，眺望老城厢，还能欣赏到传统曲艺表演。

The Haishang-liyuan is veiled on the 4th floor of the Hefeng Building on Wenchang Road. From an outdoor platform, one can overlook Yuyuan Garden and the Old Town while enjoy the shows of traditional folk art.

建筑师	禾诣轩设计
建成时间	2016 年
建筑功能	曲艺表演
建筑高度	16.5 米
建筑层数	4 层
地址	文昌路 10 号
开放时间	09:30—18:00
公共交通	地铁 10、14 号线豫园站；公交 66 路河南南路福佑路站，公交 11、64 路新北门站
Architect	Heyixuan Design
Built	2016
Building features	Chinese Folk Art Performance
Total building height	16.5m
Total floors	4
Address	10 Wenchang Rd.
Recommended viewing time	09:30-18:00
Public Transportation	Metro Line 10/14 Yuyuan Garden Station; Bus 66 Henan Rd.(S)/Fuyou Rd. Station; Bus 11/64 Xinbeimen Station

豫园地区
Yuyuan Garden Area

小南门火警钟楼
The Fire Alarm Bell Tower in Xiaonanmen

老城厢，指黄浦区由人民路、中华路围合的约 2 平方公里区域，被称为上海城市发展的"根脉"，是上海历史的发祥地。它大约成陆于 1700 年前，11 世纪逐渐发展成为海运贸易港口的集镇。南宋年间（13 世纪中后叶）上海镇正式设立，并且设置市舶提举司及榷货场，负责检查出入海港的船舶以及征收商税，镇署在今旧校场路一带。元二十八年（1291）上海正式设县。明嘉靖三十二年（1553），为抵御倭寇入侵，一座周长 4.5 公里的圆环形城墙在三个月内完工，墙内称"城"，墙外称"厢"，共筑有 6 座城门，分别是：朝宗门（大东门）、跨龙门（大南门）、仪凤门（老西门）、晏海门（老北门）、宝带门（小东门）、朝阳门（小南门），其中朝宗、仪凤和宝带 3 座城门并设水门（朝阳门后增设水门）。

开埠后的上海一度出现两个相对独立的区域——租界和华界，彼时老城厢所属的华界城区发展相对缓慢。1853 年小刀会起义军攻占上海县城，大批华人涌入租界避难，由此改变了"华洋分处"的状态。辛亥革命后，老城厢开始逐步拆除城墙、填平河浜、筑造道路，促成近代上海由传统水乡城镇向现代摩登城市转变。原先的城濠填埋后修筑环城圆路，即延续至今的人民路和中华路。1962 年开通运行的 11 路环城电车延续原先城门的名字作为站名。在 20 世纪 40—80 年代，老城厢除了少量公共设施建设，以及 1958 年开辟河南南路，城市形态并未发生太大变化，大境阁下和露香园路旁未被拆除的两段明代城墙亦得以保留至今。

1990 年代，随着复兴东路隧道建设和道路拓宽，老城厢由原先完整的圆形区域，转变为由河南中路与复兴东路两条城市主干道分隔的四个"象限"。沉浮俯仰之间，城隍庙、沉香阁（又叫慈云寺）、文庙，以及小南门火警钟楼等散落的历史地标串联起老城厢七百多年的发展故事。

The Old Town, which refers to the area about 2 square kilometers enclosed by Renmin Road and Zhonghua Road in Huangpu District, is known as the "root of Shanghai." It is the birthplace of urban Shanghai. It was formed about 1,700 years ago. During the 11th century, it gradually had developed into a shipping port and trade market. In the Southern Song Dynasty (around the middle of the 13th century), along with the official establishment of Shanghai Town, a shipyard was set up to inspect ships entering and leaving the harbor and to collect business taxes. The town hall was near the Jiu Jiaochang Road. In 1291, the Shanghai County was sited formally. In 1553, in order to defend the Japanese pirates, Shanghai County began to build a block-out wall. In three months, a 4.5–kilometer of circular ramparts was completed. The area inside the wall was called "the city or town" while the outside was called "the wing." There were six gates around the ramparts: the Dangdongmen (Great East Gate), the Dananmen (Great South Gate), the Laoximen (Old West Gate), the Laobeimen (Old North Gate), the Xiaodongmen (Little East Gate), the Xiaonanmen (Little South Gate). To reinforce the defensive power, moats were also put up at the Dadongmen, the Laoximen, and the Xiaodongmen. Later, a moat was added at the Xiaonanmen.

There existed two relatively independent regions in Shanghai for a period after the port opening: Chinese territories (Old Town) and the Concessions. Comparing to the concessions, the development of the Chinese community in the Old Town area was relatively slow. In 1853, a rebellion group "the Swords Society" attacked and seized the Old Town. Many Chinese came into the concessions to seek asylum. The state of the separation between Chinese and foreigners therefore changed. After Xinhai Revolution, the reconstructions of the Old Town started undergoing: some walls were torn down, some creeks and rivers were filled, and roads were built. With the gradual disappearance of the traditional "water town," Shanghai began to approach the modern era. Today's Renming Road and Zhonghua Road are the descendants of the loop around the city after the moats were filled.

The appearance of the Old Town didn't have much change during the 1940s and 1980s, except a couple of public service facilities and South Henan Road built in 1958. All stations on the route of tram 11 (1962) were named after the original town gates. Two small sections of ramparts from the old city wall near the Dajingge Daoist Temple and the Luxiangyuan Road have been preserved to this day.

In the 1990s, after East Fuxing Road was widened and the East Fuxing Road Tunnel was constructed, the Old Town was divided by Middle Henan Road and East Fuxing Road into four sub-regions just like four "quadrants" inrespect of two axes. The Chenghuangmiao Temple, the Chenxiangge Monastery (also called Ciyunsi Temple), Confucius' Temple, and the Xiaonanmen Bell Tower are like jewel beads on a silver string refracting the unforgettable history of urban Shanghai.

大境阁
Dajingge Taoist Temple

白云观
Baiyunguan Taoist Temple

豫园商城 | Yuyuan Bazaar

豫园商城

商城毗邻豫园和城隍庙,占地面积5.3万平方米,起源于清同治年间的老城隍庙市场,集邑庙、园林、建筑、商铺、美食、旅游等于一体,具有浓郁的民俗风情。1987年成立上海首家股份制试点企业,作为"中华商业第一股"于1992年正式上市。从1991年开始,豫园商城经过几次大规模改扩建,形成明清风格仿古建筑群。2016年至今完成新一轮改造,以商业空间围合中心景观、退墙筑廊、开放建筑界面,形成立体的公共空间与商业动线。

Yuyuan Bazaar

Originated from the Old Town's bazaar in the Qing Dynasty, Yuyuan Bazaar is adjacent to Yuyuan Garden and Chenghuangmiao Temple, covering a geographic area of 53,000 square meters. Full of folk customs, it is an interesting and attractive composition of worshipping temples, landscape gardens, architectures, businesses, food, and tours. The first firm with joint stock system was founded here in 1987. The stock entered in the market in 1992. Since 2016, Yuyuan Bazaar has been completed a new round of reform, which had taken on a new look of multi-dimensional public space and dynamic commercial environment.

30F
开放空间
Open Space

外滩英迪格酒店
Hotel Indigo on the Bund

外滩英迪格酒店：黄浦江跨江系统
Hotel Indigo on the Bund: The Cross-river System of Huangpu River

位于十六铺旅游中心旁，酒店顶层为退台式设计，位于三十层的酒吧室外平台可以环顾浦江两岸风光。

The Hotel Indigo on the Bund is located next to the current Shiliupu Tourism Center. The top floor of the hotel has been recessed, so that the CHAR bar on the 30th floor has an outdoor platform overlooking the scenery on both sides of the Huangpu River.

建筑师	赫希贝德纳联合设计顾问公司
建成时间	2010 年
建筑功能	酒店
建筑高度	114.6 米
建筑层数	31 层
地址	中山东二路 585 号
开放时间	13:00—次日 01:00
公共交通	地铁 10、14 号线豫园站；公交 55、65 路十六铺站，公交 11、64 路小东门站
Architect	Hirsch Bedner Associates (HBA)
Built	2010
Building features	Hotel
Total building height	114.6m
Total floors	31
Address	585 Zhongshan Rd. (E-2)
Recommended viewing time	13:00-01:00 (next day)
Public Transportation	Metro Line 10/14 Yuyuan Garden Station; Bus 55/65 Shiliupu Station; Bus 11/64 Xiaodongmen Station

近代兴建的交通基础设施对上海的发展产生了深远影响，并已成为重要的城市文脉。紧邻老城厢东侧的滨江区域十六铺码头，曾经是老上海重要的码头渡口，如今更是滨江游玩的好去处。在临江的英迪格酒店露台上，可以俯瞰游船如梭的黄浦江。轮渡作为联系黄浦江两岸的交通工具，可以追溯到1911年。地方自治机构"塘工善后局"租赁小火轮行驶于黄浦江东西两岸，载运旅客、酌收渡费，这便是官办黄浦江轮渡的开始。第一条轮渡航线为东铜线，从浦西铜人码头（今南京东路外滩）直航浦东东沟码头。1947年，官商合办成立上海市轮渡公司，形成以市轮渡为龙头、民营济渡为补充的格局，为此后上海渡运业奠定了基础。1970年代之前，市民过江只能依靠摆渡。

1971年通车的打浦路隧道不仅是上海的第一条越江隧道，也是中国的第一条越江隧道；1975年和1976年，松浦大桥建成双层铁路公里两用桥，改写了黄浦江无大桥的历史。改革开放后，黄浦跨江系统跟随城市总体规划，从1988年的延安东路隧道北线和1991年的南浦大桥开始，跨江系统的建设速度和技术创新都创造了前所未有的记录。

如今外滩两岸跨江系统由三部分构成。一是轮渡系统，包括浦江沿线的机动车轮渡、客轮渡和自行车专渡。截至2021年，黄浦江上运营的轮渡路线共有17条。二是越江工程，包括13座跨江大桥（含一座铁路桥）、19条车行隧道（含2条在建隧道）、1条外滩人行观光隧道以及多达17条（含在建）跨江轨道交通。三是与越江工程和轮渡设施配套的公交线路。

黄浦江全长约113公里，通过跨江系统的建设，将浦东与浦西紧密相连，改变了"宁要浦西一张床，不要浦东一间房"的老观念。2002年启动的浦江两岸综合开发计划，长约42.5公里，包含从吴淞口到徐浦大桥的区域，被认为是继浦东开发以来上海第二个里程碑事件。浦江两岸综合开发，推动上海成为全国最大的经济、贸易、科技、金融、信息中心。

十六铺码头与古城公园
Shiliupu Pier and Gucheng Park

The construction of modern transportation infrastructure has had a profound impact on Shanghai's development and has become an important urban context.

To the east of the Old Town, just along the riverside, the former Shiliupu Pier was once an important ferry pier in Shanghai. Now it has become a viewpoint for river sightseeing. On the terrace of the Hotel Indigo on the Bund, we can overlook the scene of cruises and ferries on the Huangpu River. As a method of the transportation connecting the two sides of the Huangpu River, ferry service can be traced back to 1911. The "Tanggong Rehabilitation Bureau" of Shanghai local autonomous organization rented a small vessel ferry to carry people back and forth. The first ferry route was Dongtong Route from Tongren Pier in Puxi (now the Bund of East Nanjing Road) to Donggou Pier in Pudong. In 1947, Shanghai Ferry Company was established. It set up the ferry transportation system in which municipal ferry was dominant and the private ferry was auxiliary. This system had laid the foundation of ferry service for the future in Shanghai. Before the 1970s, ferries were the only transportation service crossing the river.

Built in 1971, Dapu Road Tunnel was the first cross-river tunnel in Shangghai as well as in China. Songpu Bridge, the first cross-river railway bridge and cross-river road bridge was completed in succession in 1975 and 1976. With the governmental policy of Reform and Opening-up, the overall urban planning was carried out. Following the completion of the East Yan'an Road Tunnel in 1988 and Nanpu Bridge in 1991, the construction of cross-river system has been sped up and the advanced constructional techniques have been implanted.

Today, the cross-river system on both sides consists of three parts. The first part is the ferry system which combines motor ferries, passenger ferries, and bicycle ferries along the Huangpu River. Before 2021, there were 17 operating ferry lines. The second part is the cross-river projects, including 13 cross-river bridges (containing 1 railway bridge), 19 cross-river tunnels (include 2 tunnels being built), 1 sightseeing tunnel on the Bund, and more than 17 metro lines. The third part is the multiple bus lines supporting the cross-river project and ferry facilities.

The Huangpu River is about 113 kilometers long. Cross-river transportation system connects the life on two shores: Pudong and Puxi. Launched in 2002, the Huangpu River Comprehensive Development Plan broke the very lasting cogitation:"it'd be better to have a bed in Puxi rather than owning a home studio in Pudong." This plan vastly covered the area from Wusongkou to Xupu Bridge with linear distance of 42.5 kilometers. It is undoubtedly considered to be the second milestone in Shanghai since the exploitation in Pudong. The comprehensive development of both sides of the Huangpu River has promoted Shanghai to become the center of commerce, trade, science and technology, financial, and information in China.

黄浦江畔的十六铺旅游码头
Shiliupu Pier by the Huangpu River

十六铺码头

曾经是上海的"水上门户"。早在上海开埠前,十六铺就是东亚最大的码头。开埠后上海的外来文化与老城厢民俗文化相结合,形成十六铺地区独有的"码头文化"。中华人民共和国成立之后,这里一直是上海港最主要的客运码头。随着城市的整体规划调整,2003年,十六铺码头最后一条定期客运航线搬迁至吴淞客运中心。从2006年开始,分期建设与改造的"新十六铺",成为集邮轮码头、社会停车等功能于一身的新地标。

Shiliupu Pier

The famous Shiliupu Pier in Shanghai was the largest pier in China and East Asia in the Qing Dynasty. Before 1949, it was used as a merchants and trading channel. It had been a key ferry terminal port in Shanghai by the end of the 20th century. Later, with the overall planning of the city, the pier was transformed into a complex of waterfront scenery landscape, a broad-scaled port, a comprehensive commercial center, social spots, parking space, and other beneficial functions. It has become Shanghai's new landmark with historical and practical significance.

摩天轮
Ferris wheel
开放空间
Open Space

大悦城摩天轮
Ferris Wheel in the Jing'an Joy City

静安大悦城：铁路建设的起点
Jing'an Joy City: The Starting Point of Railway Construction

位于苏河湾以北的静安大悦城，以其摩天轮的独特顶部空间而闻名。随着摩天轮的缓缓转动，可以放慢时间的脚步，一览上海铁路的前世今生，感受城市快速路与轨道交通的更迭发展。

Located to the north of Suhewan Zone, Jing'an Joy City is famous as its unique roof space combined with a Ferris wheel called "Sky ring." As the Ferris rises up into the air slowly, tourists can see the trains carrying thousands of passengers on the crisscross railways. The stunning changes and developments of urban rail transit in Shanghai are reflected in here.

建筑师	楷亚锐衡有限公司
建成时间	2010 年（一期），2016 年（二期）
建筑功能	零售、娱乐、文化等
建筑高度	100 米（含摩天轮）
建筑层数	11 层
地址	西藏北路 166 号
开放时间	10:00—22:00
公共交通	地铁 8、12 号线曲阜路站；公交 46 路西藏北路海宁路站，公交 15 路曲阜路西藏北路站
Architect	Callison RTKL
Built	2010/2016
Building features	Retail, Entertainment, Culture, etc
Total building height	100m (Including Ferris Wheel)
Total floors	11
Address	166 Xizang Road (N)
Recommended viewing time	10:00-22:00
Public Transportation	Metro Line 8/12 Qufu Rd. Station; Bus 46 Xizang Rd.(N)/Haining Rd. Station; Bus 15 Qufu Rd./Tibet Rd.(N) Station

静安大悦城与上海火车站
Jing'an Joy City and Shanghai Railway Station

苏州河畔西藏北路的静安大悦城，上有 56 米直径的国内首个屋顶悬臂式摩天轮。整个商业综合体以商业服务与游乐设施相结合的全新模式吸引年轻人和周边居民，与一河之隔的老牌南京东路商圈优势互补。乘坐摩天轮自北向西远眺，高楼林立之间一条铁路线横亘东西，见证了上海开埠后中国铁路一个多世纪的荣辱兴衰。

1876 年，英商怡和洋行借修路之名修建吴淞铁路，成为中国土地上最早的营业铁路。保守的清政府害怕铁路通车将导致大量百姓失业，又迫于修路强征百姓土地引发的民愤，最终赎回铁路，并于 1877 年将铁路路轨全部拆除。短暂运行的铁路所展现的高效，让清政府意识到这一新型基础设施的价值。1897 年，清政府在上海成立铁路总公司，按吴淞铁路走向重建淞沪铁路（1898），该段后成为年沪宁铁路（1908）支线。

1909 年落成的沪宁铁路上海站（1916 年改名"上海北站"）为英式古典风格建筑。在 1932 年淞沪抗战和 1937 年淞沪会战期间，连续遭遇战火，并历经重建与修复，于 1947 年升格为上海总站，下辖上海北站、麦根路货站等站点。1950 年，上海北站改名为上海站，担当起新上海的陆上运输大门，承载南来北往客，成为老百姓记忆中亲切的"老北站"。

随着经济的发展，老北站逐渐无法满足上海陆上运输的庞大吞吐量。1984 年，市政府决定在原铁路上海东站（麦根路货站）的基础上兴建上海铁路新客站（现上海站）。老北站于 1987 年新客站启用之时停止服务，2004 年在原址上建成上海铁路博物馆。淞沪铁路废弃后，沿原线路建成高架轻轨，2000 年投入使用，服务于地铁 3 号线、4 号线。曾经贯穿上海的重要铁路线虽然消失，但轨道运输的传统和周边的城市文脉延续至今，并成为俯瞰视野中鲜明而强烈的城市空间要素。

静安大悦城顶层小街区商业
Small-block on the top floor of Jing'an Joy City

On the north bank of the Suzhou Creek, Jing'an Joy City has the first cantilever Ferris wheel of 56 meters in diameter on rooftop in China. By combining businesses and amusement facilities with a new model for young customers, Joy City can compete with the East Nanjing Road business district across the Suzhou Creek. Looking to the northwest from the rotating Ferris wheel, we can see a railway line straddling across the east and the west between skyscrapers and witness the history and development of railroad in China for more than a century.

In 1876, the Shanghai-based British merchant Jardine Matheson built the Wusong Railway under the guise of repairing roads. Fearing the public anger from residents losing jobs and lands caused by the railway constructions, the conservative Qing government had several disputes with the merchants and managed to purchase the railway at the original cost. This railroad was demolished in 1877. But the efficiency of rail transportation also made the Qing government realize the value of this new type of infrastructure. In 1897, the China Railway Corporation was established in Shanghai. In 1898, the new Wusong Railway was built, later became a branch line of Huning Railway (1908, Shanghai-Nanjing Railway).

In 1909, Shanghai North Railway Station (formerly known as Shanghai-Nanjing Railway Shanghai Station) was officially put into use. It is the first railway station in Shanghai and a milestone in the development of Shanghai Railway. During the War of Resistance against Japanese Aggression, the North Railway Station was bombed by Japanese military attacks quite a few times. After lots of restorations and renovations, the station was upgraded and became Shanghai Terminus in 1947. The regions included Shanghai North Railway Station, Markham Road Yard, etc. In 1950, the North Railway Station renamed Shanghai Railway Station. It was the gateway of the land transportation in Shanghai. Thousands of passengers travelled between the north and the south had passed through this station. It had been a forever-lasting memory in the hearts of so many people.

With the economic development, the North Railway Station gradually had become insufficient to the huge throughput on Shanghai's land transportation. In 1984, the municipal government decided to build a new central station based on the structure of the original Shanghai East Railway Station (Markham Road Yard). The North Railway Station ended its facilitating in 1987 when the new Shanghai Railway Station opened its business. It was converted into the Shanghai Railway Museum in 2004. The earlier Songhu Railway was rebuilt as an elevated metro service. Metro Lines 3 and 4 started running along the original railway line in 2000. Although the important railway line that once ran through Shanghai disappeared from the city transportation structure, the tradition of rail transportation and the surrounding urban context continue to this day, and have become a distinct and strong spatial element, clearly visible from this overlook.

上海铁路博物馆

博物馆位于天目东路 200 号,2004 年 8 月建成开放,4 层主楼根据 1909 年沪宁铁路上海站原样,按照 80% 的比例建于原址之上。博物馆展示了从 1860 年代铁路进入中国后,上海又华东铁路一百多年来走过的历程,突出反映铁路生产力的发展。

Shanghai Railway Museum

The Museum is located at No. 200 on East Tianmu Road. Completed and opened in August 2004, this four-story main building is basically built according to the plan of the original old North Railway Station, at a rate of 80%. The museum shows visitors the course of the railways in China since the 1860s.

顶层
The top floor
开放空间
Open Space

新世界丽笙大酒店
Radisson Blu Hotel Shanghai New World

新世界丽笙大酒店：南京路的商业发展
Radisson Blu Hotel Shanghai New World: The Commercial Development of Nanjing Road

酒店顶部标志性的"飞碟"使其成为市中心的地标，位于四十五楼的旋转景观餐厅与星空酒吧，旋转一圈需要 2 小时。此处可以居高临下，观赏繁华的南京路步行街，尽览城市盛景。

There is an iconic "flying saucer" shaped revolving restaurant and "starlight bar" on the 45th floor of the Radisson Blu Hotel Shanghai New World. The complete rotation of restaurant takes two hours. It is the best place to feel the prosperity and intensity of the Nanjing Road.

建筑师	华东建筑设计研究院有限公司
建成时间	2003 年
建筑功能	酒店
建筑高度	208 米
建筑层数	47 层
地址	南京西路 88 号
开放时间	11:30—次日 02:00
公共交通	地铁 1、2、8 号线人民广场站；公交 20、37 路九江路黄河路站，公交 18、49 路人民广场（福州路）站

Architect	ECADI
Built	2003
Building features	Hotel
Total building height	208m
Total floors	47
Address	88 Nanjing Rd. (W)
Recommended viewing time	11:30-02:00 (next day)
Public Transportation	Metro Line 1/2/8 People's Square Station; Bus 20/37 Jiujiang Rd./Huanghe Rd. Station; Bus 18/49 People's Square (Fuzhou Rd.) Station

旋转餐厅的视野
View from the revolving restaurant

从静安大悦城摩天轮南望,一栋顶端如飞碟悬停的超高层建筑,分外醒目,其便是位于南京路核心地段的新世界丽笙大酒店。近代上海城市的繁华,以外滩为源头,由东向西发展,并一度在南京路形成制高点。从酒店的旋转餐厅向下俯瞰,外滩、南京东路和人民公园等尽收眼底。

南京路的兴起源于租界时期跑马场的建设。从1845年开始,英侨建造的跑马场经两次搬迁,最终于现在的人民公园处建成跑马厅。最初的跑马场(又称"花园")位于外滩附近花园弄一带(现南京东路以北、河南中路以西),并在花园南侧外增筑马道,"马路"便由此而来。为便于人们前往赛马场,租界当局修筑了一条与南侧马道平行的道路,直抵外滩,称"大马路",1865年正式定名为南京路。随着19世纪末的"英商四大百货公司"(福利、汇司、泰兴、惠罗)和20世纪以来由华商开设的"中国四大百货公司",南京路开创了亚洲百货业的诸多先河,成为名扬海外的远东第一商街。

1943年,民国政府将南京路改称南京东路、静安寺路改称为南京西路。1949年后,南京东路历经改造更新,汇集了上海乃至全国的名特优老牌产品,被誉为"中华商业第一街"。1995年,南京东路试行周末步行街制度,成为我国最早的商业步行街之一,开启了复合功能开发的序幕。1999年,秉持"以人为本"的城市设计理念,经过调整商业布局,改善街区环境,南京东路步行街全面建成并正式开街。2020年,南京东路步行街东拓至外滩,构建起外滩与南京路步行街之间一体化的步行观光购物通道,成为展现上海全球卓越城市魅力的新窗口。

Looking to the south across the Suzhou Creek from the Jing'an Joy City, you can see a high-rise building with an UFO-like object hovering the top. This is the Radisson Blu Hotel Shanghai New World at the joint of East Nanjing Road and West Nanjing Road. The prosperity of modern Shanghai began from the Bund. The continuous economic growths spread from the east side of the city towards the west side, and Nanjing Road was the most flourishing section for a time. Overlooking from the revolving restaurant in the hotel, you can have a panoramic view of East Nanjing Road from the Bund to the People's Park.

The rise of Nanjing Road originated from the construction of a racecourse during Concession. Beginning in 1845, after two relocations, a horse racing hall finally was set in the current People's Park. The original racecourse (also named Garden) was located near the Bund along Huayuan Road (now north of East Nanjing Road and west of Henan Middle Road), and a horse track was added outside the south side of the Garden, from which the "Horse Road" was derived. In order to facilitate people's access to the racecourse, the concession authorized to build a road parallel to the horse track on the south side. This road was called "Main Horse Road," reaching all the way to the Bund. It was officially named Nanjing Road in 1865. With the openings of the "Four Department Stores" by British businessmen at the end of the 19th century and the "New Four Department Stores" by Chinese businessmen in the 1920s and 1930s, Nanjing Road had become the first modern commercial street in the Far East.

In 1943, the government of the Republic of China renamed Nanjing Road to East Nanjing Road, and Jing'ansi Road to West Nanjing Road. After the founding of the People's Republic of China, East Nanjing Road had undergone transformation and upgrading. It was also known as "the First Commercial Street in China." In 1995, East Nanjing Road was implemented as a pedestrians-only street without any motor vehicles on the weekends. This sped up the development of complex commercial functions. In 1999, the reconstruction project of East Nanjing Road pedestrian street, was successfully completed. The project revised and regulated the earlier commercial function structure, and improved the overall environment of the block based on "people-oriented" urban design. In 2020, the municipal government completed the reconstruction of East Nanjing Road Pedestrian Street to the east, which is connecting with the Bund, giving this time-honored commercial street a new life.

新世界丽笙大酒店 | Radisson Blu Hotel Shanghai New World

新世界丽笙大酒店与人民广场
Radisson Blu Hotel Shanghai New World and the People's Square

四大百货公司

20世纪初,随着中国民族工商业的快速发展以及西方商业的引入,综合性的商业模式开始在中国出现。南京路"四大百货公司",包括先施公司(1917,现上海时装商城、东亚饭店)、永安公司(1918,现永安百货)、新新公司(1926,现上海第一食品公司)、大新公司(1936,现上海市第一百货商店)。它们集商业、住宿、餐饮、金融、文化体育和娱乐于一体,开创了中国近代商业建筑的先河,出现了首座自动扶梯、空中花园等,也是上海现代化城市进程的重要载体。永安百货屋顶上的绮云阁还是上海解放时第一面红旗升起的地方。

The Four Department Stores

The Four Department Stores on Nanjing Road, are referred to Sincere Company (1917, now Shanghai Fashion Store, East Asia Hotel), Wing On Department Store Company (1918, now Yong'an Department Store), Sun Sun Company (1926, now Shanghai First Foodhall) and the Sun Company (1936, now Shanghai No.1 Shopping Center). In the 1920s and 1930s, the rise of the four major department stores made Nanjing Road a world-class commercial street. The very first escalator appearing in the entire country, sky gardens, financial institutes, restaurants and lodges, public entertainments, and all other facilities and businesses existed in those stores initiated the modern model of commerce complex.

上海市第一食品商店
Shanghai First Foodhall

上海时装商城
Shanghai Fashion Store

上海第一百货
Shanghai No.1 Shopping Center

永安百货
Yong'an Department Store

美罗城：
徐家汇的变迁
Metro City:
Changes of
Xujiahui

美罗城的巨形球体是徐家汇的地标之一，其二层衔接徐家汇空中连廊，可往来徐家汇地区的各大商业楼宇，沿途有丰富的景观、绿化和休憩设施，空中漫步，纵览徐家汇全景。

The huge globe of Metro City is an iconic landmark of Xujiahui. From the second floor of Metro City, you can walk to the Skyway where the platform connects many important buildings in Xujiahui, with rich green and relaxing facilities available for citizens all day.

建筑师	新加坡 BJ 建筑设计事务所 华东建筑设计研究院有限公司
建成时间	1997 年
建筑功能	零售、剧场
建筑高度	41.2 米
建筑层数	8 层
地址	肇嘉浜路 1111 号
开放时间	10:00—22:00
公共交通	地铁 1、9、11 号线徐家汇站；公交 42、43 路徐家汇站，公交 15、44 路天钥桥路辛耕路（徐家汇）站
Architect	Singapore BJ Architects International, ECADI
Built	1997
Building features	Retail, Theater
Total building height	41.2m
Total floors	8
Address	1111 Zhaojiabang Road
Recommended viewing time	10:00-22:00
Public Transportation	Metro Line 1/2/9 Xujiahui Station; Bus 42/43 Xujiahui Station; Bus 15/44 Tianyaoqiao Rd./Xingeng Rd. (Xujiahui) Station

徐家汇与八万人体育场
Xujiahui and Shanghai Stadium

徐家汇被称为"海派文化之源",这里既是中国近代最具规模与影响力的中西文化与科学交流地,也见证了中国电影和唱片业的发展繁荣。

徐家汇源于中国近代科学先驱徐光启(1562—1633,中国最早的天主教徒之一)及族人在此繁衍生息。徐家汇初名徐家库(音 shè),19 世纪上半叶改称徐镇,原先是蒲汇塘、肇嘉浜和李漎泾三条河流的汇合处,故名"徐家汇"。鸦片战争以后,法国天主教传教士进入上海。1847 年,江南教区选择水上交通便捷、具有天主教传统的徐家汇建造耶稣会会院及一系列附属设施,其中徐家汇观象台(1873,现上海气象博物馆)是中国沿海第一座观象台,徐家汇藏书楼(1897)是上海现存最早的近代图书馆,圣依纳爵公学(1850 年,也称徐汇公学,现徐汇中学崇思楼一楼汇学博物馆)是中国最早按西洋办学模式设立的学校之一,土山湾画馆和土山湾孤儿院(1852,1864,现为土山湾博物馆)不仅是中国西洋画的摇篮,也是近代新工艺、新技术的发源地,在近代中西文化交流中起到举足轻重的作用。1910 年,哥特式建筑风格的圣依纳爵堂(现徐家汇天主堂)建成,徐家汇成为上海天主教活动中心之一。

1930 年代前后,逃荒来沪的苏北多地农民在肇嘉浜两岸搭棚落脚谋生。1937 年"八一三"事变,难民再次蜂拥至此,形成全城最大的水畔棚户区和"臭水浜"。彼时海格路(华山路)以东的贝当路(衡山路)、姚主教路(天平路)一带多为高级住宅、公寓和新式里弄,当局架设隔离网阻隔难民进入。属于华界的徐家汇地区,在徐镇路(虹桥路近徐家汇)一带设立临时难民点,形成人口稠密的棚户区。1954 年,上海开始治理肇嘉浜,填浜筑路,并形成一条宽阔的绿化带,成为西南区域重要的交通大动脉。1984 年,徐家汇人行天桥建成。1993 年,地铁 1 号线南段试运行从徐家汇站起,围绕轨交站建造了一批百货商店。2001 年在大中华橡胶厂和中国唱片厂原址(百代唱片公司旧址)建成徐家汇公园。2013 年上海电影博物馆落户上海电影制片厂原址。

如今,徐家汇地下已有 18 个出入口的三线轨交换乘站。2022 年竣工的空中连廊二期,在为市民提供沉浸式购物体验的同时,串联周边众多"海派之源"文化遗存,打造了具有鲜明识别度和活力的城市副中心。

Xujiahui, known as "the springhead of Shanghai culture," is the largest and most influential cultural and scientific exchange place between China and the West in modern years. It was once the center of Shanghai's film industry and phonogram industry.

The formation of this area can be traced back to the Ming Dynasty, when Paul Xu (Xu Guangqi, 1562-1633, known as the pioneer of modern

Chinese science, and was also one of the earliest Catholics in China) and his race peopled here. It was firstly named "Xujiashe", and renamed Xu Town in the first half of 19th century. It is the place of three rivers met, also named "Xujiahui." (Chinese"hui"means met)

French Catholic priests came to Shanghai after the Opium War. In 1847, the Catholic Jiangnan diocese chose Xujiajhui, in consideration of its convenient water transport and Catholic tradition, to build the Jesuit Catholic Church and a series of ancillary buildings. Among them, the Zi-Ka-Wei Observatory (1873, now Shanghai Meteorological Museum) is the first observatory along the coast of China, and the Xujiahui Bibliotheca is the earliest modern library in Shanghai. College of St. Ignace (now Xuhui High School, which Chongsi Building's first floor is the Huixue Museum today) was one of the first schools in China in accordance with the Western school model. Orphelinat de T'ou-Sè-Wè and T'ou-Sè-Wè Studio are not only the cradle of Chinese Western painting, but also the birthplace of modern new crafts and new technologies. They played an important role in modern cultural communication between China and the West. In 1910, the Gothic architectural style of the St. Ignatius Cathedral (now Xujiahui Catholic Church) was completed, and Xujiahui became one of the Catholic activity centers in Shanghai.

Around the 1930s, floods and famine victims flee from outside Shanghai and built shelters along the banks of Zhaojiabang River to make a living on. On August 13, 1937, Japanese army entered the Chinese border in Shanghai, and thousands of more refugees were packed into the over-crowded river bank, which later became the largest "shanty town"and "smelly water." Back then, local authorities set fences to prevent refugees into the neighboring French Concession, and resettled them at Xuzhen Road (now Hongqiao Road near Xujiahui) and surroundings, gradually forming a populated slum area.

In 1954, the municipal government began to filling Zhaojiabang River and paved a boulevard which became a traffic artery in the southwest of Shanghai. In 1984, an overpass was built in the center of Xujiahui. In 1993, the southern section of Metro Line 1 was put into trial running, bringing about a number of commercial buildings. In 2001, Xujiahui Park was opened on the former site of Shanghai Ta Chung Hua Rubber Factory and Shanghai China Record Co. (Pathe Villa is the recording site of the March of the Volunteers). In 2013, Shanghai Film Museum was opened on the site of Shanghai Film Studio.

Till now, Xujiahui has become a transfer center of three metro lines with 18 exits and a skyway was completed in 2022. It offers a better business environment and more shopping fun while connecting many cultural remains of the Shanghai style in the surroundings.

美罗城全景
Panoramic view of Metro City

- 虹桥商务区 Hongqiao CBD
- 徐家汇中心 International Trade Center
- 上海西藏大厦万怡酒店 Courtyard by Marriott Shanghai Xujiahui
- 中国科学院上海天文台 Shanghai Astronomical Observatory, Chinese Academy of Sciences
- 光启公园(徐光启纪念馆) Guangqi Park (Xu Guangqi Memorial Hall)
- 徐家汇观象台 The Zi-Ka-Wei Observatory
- 华亭宾馆 Huating Hotel
- 地铁3号线 Metro line 3
- 上海体育场(八万人体育场) Shanghai Stadium
- 徐浦大桥 Xupu Bridge
- 龙华烈士纪念馆 Longhua Martyrs Memorial
- 港汇恒隆广场 Grand Gateway
- 徐汇中学(原徐汇公学旧址) Shanghai Xuhui High School (former site of College of St Ignatius)
- 圣依纳爵主教座堂 St. Ignatius Cathedral
- 徐家汇圣母院旧址 Former site of Notre Dame de ZI-KA-WEI
- 太平洋百货 Pacific Department Store

徐家汇藏书楼

位于漕溪北路80号,始建于1847年,是上海现存最早的近代图书馆。初为3间用以储藏图书和档案的"修士室",后经1860年和1897年两次扩建形成独立的双层藏书楼,上层为西文书库,布局和藏书架为梵蒂冈图书馆式样;下层为中文书库,仿明代宁波天一阁,建筑也体现了中西合璧的特色。1956年并入上海图书馆,近年来几经修葺,开放程度不断提高,并以中西书籍藏品的丰富与珍贵闻名于世。

Xujiahui Bibliotheca

Xujiahui Bibliotheca (known as Bibliotheca Zi-Ka-Wei), located at No. 80 on North Caoxi Road, founded in 1847, is the earliest existing modern library in Shanghai. This 3-room library, where books and archives were stored, was first built in Xujiahui by the Shanghai Catholic Jesuit Church. These three rooms had been turned into a substantive library after two expansions in 1860 and 1897. There were two stories in the library. The top floor, replicated the styles and layouts of Vatican Library, mainly stored books in foreign languages. The bottom floor, in the fashion of the Tianyige Pavilion at Ningbo during the Ming Dynasty, was the section for Chinese books. It combined both western and Chinese traditional architectural designs. Bibliotheca Zi-Ka-Wei was incorporated into the Shanghai Library in 1956. It was claimed as a "Heritage Architecture" in Shanghai in 1994 and reopened to the public in July 2003 after renovation.

第 2 章
功能转型时期的城市空间演变

1949 年至 1990 年是上海城市发展的转型时期。落成于 1988 年的新锦江大酒店以及随后出现的明天广场、长宁龙之梦、斯格威铂尔曼大酒店、芮欧百货、上海环球港等建筑的空中开放空间，提供了观摩这一时期城市建设成果的不同视角，可感受其对当下城市格局的影响。

1949 年是上海城市发展史上的又一个重要转折点，从曾经的远东国际金融中心向社会主义工业化、现代化的城市转型，上海完成了从消费型城市向生产型城市的转变。初期艰难的国内经济形势，决定了上海在"全国一盘棋"的大局中，承担起东南沿海乃至全中国最重要的工业城市角色。与建设生产型城市相适应，从 1950 年代起，上海建设了一大批工人新村，形成颇具时代特征的城市肌理。此外，对市区现存老建筑进行功能置换与改造，大量私人园林、公墓和娱乐设施转变为重要的城市公共空间和文化设施，并提出建设近郊工业区和卫星城的规划构想。1980 年代之前，上海高层建筑建设基本停滞，市区天际线的高点是 1974 年建成的 210 米高的上海电视塔（1998 年拆除）。

改革开放后，上海重新确立以经济建设为主要目标的城市发展策略，城市迎来新的转型。首先因外事接待和旅游业发展，开始建造高层涉外宾馆。其中 1983 年开业的上海宾馆刷新了国际饭店保持将近五十年的上海第一高楼记录，1985 年落成的联谊大厦是上海第一个高度超过百米的高层办公楼，同时期建造的雁荡公寓是 1949 年后上海首幢提供涉外租住的高级公寓。

商业开发带来的土地升值使得高层建筑再次成为城市建设的主要方式。1988 年的新锦江大酒店拥有当时远东地区最大的旋转餐厅，1989 年的华东电力大楼（现艾迪逊酒店）成为 1949 年后南京路上的第一栋高层建筑，1990 年的上海商城以一组裙楼和三幢塔楼（主楼高 164.5 米）再度登上城市新高。上海终于在"一年一个样、三年大变样"的憧憬中迎来 90 年代新的建设高潮。

这一时期，城市的路网和肌理基本延续近代时期的特征，在经历社会主义改造后，诞生了一批城市公共空间和公共设施。它们成为城市重要的空间节点，并深刻地改变了市民的生活方式，成为许多人对上海的独特记忆。

CHAPTER 2
The Evolution of Urban Space during the Transformation Period

The transition period of the urban development in Shanghai was between 1949 and 1990. Tomorrow Square, Cloud Nine, Pullman Shanghai Skyway Hotel, Réel Department Store, Global Harbor following Jin Jiang Tower Hotel, built in 1988, all feature open-air social facilities. These facilities, open to the public, provide visitors different aspects to cherish the urban construction achievements and to sense the impact on the current pattern in urbanization.

1949 was another significant turning point in the history of urban development in Shanghai. The transition of urban lifestyle from consumption to production was completed in that time period. Shanghai, once was the center of commerce in the Far East, had become an industrial city and started to accelerate its modernization. In the early days after the founding of the People's Republic of China, the nation was facing all kinds of difficulties in building up the economy. With this background and following the government policy, "taking the whole country into one account," Shanghai played an important role in industry, undertaking large and heavy industrial tasks and providing most of the needs to the regions along the southeast coast, even to the entire country. Adapting such a transition, many residential buildings started to stand upright from the 1950s, homing manufacturing workers. Those large-scaled neighborhoods had formed urban living style and architectural diagram with characteristics of the times. Meanwhile, renovation works had taken place. Old buildings, private gardens, cemeteries, and entertainment arenas had been replaced into public socializing platforms. Industrial districts and their surrounding satellite towns already had been put in the plans. Before the 1980s, the constructions of high-rising buildings stagnated. The 210-meter-tall Shanghai TV Tower (1974, demolished in 1998) reached the highest point of downtown skyline.

As the Reform and Opening-up policy was carried out, the schemes of urban development in Shanghai were reestablished. All the plans focused on one main goal: economic development. What appeared first were the tall and fancy hotels lodging tourists mostly from overseas. The height of Shanghai Hotel, opened its business in 1983, broke the record of Park Hotel having been the tallest building in Shanghai for neary a half century. In 1985, two constructions brought the nation's attentions: one is the Union Building. It is the first building with its height over 100 meters. Another one is Yandang Buiding. It is the first high-class apartment complex since 1949 welcoming the foreign renters.

The land appreciation caused by continuous commercial developments brought the urban constructions into a new yet necessary structure: advantage the upper air space. Jin Jiang Tower Hotel, built in 1988, featured the largest revolving restaurant in the Far East at the time. East China Electric Power Building (now Shanghai EDITION), built in 1989, had become the first high-rising building on the Nanjing Road after 1949. In 1990, a set of podiums and three towers (the height of main tower is 164.5 meters) in Shanghai Center had reached the city's highest point. All of these implied that Shanghai had entered a new era of architecture. The expectation of "One alteration for a year and a fresh appearance in three years" had become a joyful reality in the 1990s.

In this time period, the characteristics of the road transportation network from the old days were carried over; however, a few of newly built public facilities started to change the city's lifestyle and fulfill people's social needs.

38F~40F
开放空间
Open Space

明天广场
Tomorrow Square

明天广场：
人民广场的改造
Tomorrow Square:
The Transformation of People's Square

明天广场的尖顶塔楼是整座建筑的焦点，大厦的三十八至四十层是万豪酒店的大堂咖啡吧、中餐厅、酒廊及行政图书馆。从这里俯瞰人民广场，优越的生态环境、巨大的开放空间与周边新老建筑互相映衬，蕴藏了人民之城的时代记忆。

The tower with the tall steeple is the focal point of the entire structure. A coffee lounge, a Chinese restaurant, a bar, and an administrative library occupy from the 38th to 40th floors in JW Marriott Hotel Shanghai at Tomorrow Square.The fine ecosystem and the grand open space along with old and new architectures can be viewed from either floor. So much memories from time to time are held in the city of its people.

建筑师	约翰·波特曼建筑设计事务所
建成时间	2003 年
建筑功能	酒店、办公、零售
建筑高度	284 米
建筑层数	60 层
地址	南京西路 399 号
开放时间	07:00—次日 02:30
公共交通	地铁 1、2、8 号线人民广场站；公交 20、37 路南京西路黄河路站，公交 49 路威海路黄陂北路站
Architect	John Portman & Associates
Built	2003
Building features	Hotel, Office, Retail
Total building height	284m
Total floors	60
Address	399 West Nanjing Road
Recommended viewing time	07:00-02:30 (next day)
Public Transportation	Metro Line 1/2/8 People's Square Station; Bus 20/37 Nanjing Rd.(W)/Huanghe Rd. Station; Bus 49 Weihai Rd./Huangpi Rd.(N) Station

明天广场与人民广场
Tomorrow Square and People's Square

上海开埠后外侨们集资圈地，由东向西、三次移址扩建跑马场，带动了南京路从洋行聚集的外滩不断地向西拓宽与延伸。人民广场的前身是始建于1861年的第三跑马场。20世纪二、三十年代，在跑马总会大楼建设同期，跑马场北面陆续建成金门大酒店（1926）、西侨青年会大楼（1932，现体育大厦、体育博物馆）、国际饭店（1934）等高层建筑。

1950年和1951年，在原跑马场举办了华东区农业展览会和规模盛大的"上海市土产展览交流大会"，是解放初期疏通城乡经济血脉的重要活动。1951年，市政府决定将跑马场改造为人民广场和人民公园，修建横贯东西的人民大道、检阅台等。1952年人民公园建成后，迅速成为上海市民的娱乐休闲中心，原跑马总会大楼先后用作上海博物馆、上海图书馆、上海美术馆、上海历史博物馆等。1957年，人民公园西南角原跑马厅看台等改建为上海市体育宫（现上海大剧院）；1963年建成市人大常委会办公楼（现人民大厦）。1980年代以前，人民公园和人民广场一直是上海市民游园、集会和举行大型活动的场所。

改革开放后，城市面貌日新月异。1987年人民广场西南建成123.4米高的上海电信大楼，成为当时上海建筑的新高。1987年和1990年分别出台《人民广场地面和地下综合规划方案》《人民广场公园地区规划》，环人民广场圈改造再次被列为市委市政府重大工程。1992年，伴随地铁站建设，人民广场综合改造工程启动。经过十多年建设，人民广场从原先的硬地广场升级成以绿化为主、整体对称的开放式园林广场，保持原有广场和公园边界的同时，中轴线上分别是上海博物馆（1996）、"浦江之光"中心广场、上海市公路零公里标志、人民大厦（1995年名政府大厦，1997年更名），上海大剧院（1998）和上海城市规划展示馆（2000）位于东西两端，还有绵亘地下与地铁站相连的地下商城、亚洲最大的城市地下变电站以及上海最大的地下停车库。如今，全新的行政、文化、生态、交通、商业等功能让人民广场成为上海这座海纳百川城市的多元复合地标。

After the opening of Shanghai Port, Chinese merchants from overseas started to raise funds for purchasing more lands and expand their territories. The Horse Racing Track was relocated three times. This movement successfully extended Nanjing Road from the Bund which was a heavy cluster of commerce westward. The People's Square was once the third location of the Horse Racing Track set in 1861. Between the 1920s and 1930s, as Shanghai Race Club under the construction, Pacific Hotel (1926), Foreign YMCA Building (1932, now Sports Building & Shanghai Sports Museum), Park Hotel (1934) were built up one after another to the north realm of the racing track.

In 1950 and 1951, the Horse Racing Track was used an arena for the East China Agricultural Exhibition and the Produce Trading Fair of Shanghai. Both activities mediated and encouraged the commercial communication and exchanges between the urban cities and the countryside. In 1951, a plan to transform the racing track to the People's Square and the People's Park was determined. The People's Square crosses through east to west and later has been used for large gatherings, while the People's Park quickly had become residents' favorite leisure place. Shanghai Museum, Shanghai Library, Shanghai Art Museum, and Shanghai History Museum, these public facilities replaced Shanghai Race Club. In 1957, Shanghai Sports Palace (the site is now Shanghai Grand Theatre) was built over the original audience stand of the racing track. It was just adjacent to the southwest corner of the People's Park.

As the Reform and Opening-up policy was speeding up the development of the urban structure, Shanghai Telecom Building was put up to the southwest of the People's Square in 1987. It was the tallest building with the height of 123.4 meters in Shanghai. Two construction plans of People's Square & People's Park were introduced in 1987 and 1990. To redesign, refashion, and recast the appearance around the Square had been put on the top priority list, a comprehensive transformation started in 1992. With more than ten years' efforts, the People's Square stood as an open landscape garden filled with kinds of trees and flowers of the seasons. With the borderlines kept between the Square and the Park, this green land was symmetrical by a major axis, on which sit Shanghai Museum (1996), the fountain named as "the Light of Huangpu River," Zero Point of Highway Shanghai, People's Mansion (1995, the Building of Shanghai Municipal People's Government, 1997 renamed), Shanghai Theatre (1998) and Shanghai Urban Planning Exhibition Center (2000), the shopping mall, the largest underground urban substation in Asia, and the parking garage underground are connected by multiple routes of subway. Nowadays, the People's Square is an enormous functional complex combined with administrations, culture and art, transportations, ecological managements, and commerce. No one can refuse to acknowledge this landmark in Shanghai.

上海历史博物馆

位于南京西路325号,其前身是建于1933年的跑马总会大楼,兼具古典主义和折衷主义风格。1952年起作为上海博物馆(1959年迁出)和上海图书馆(1997年迁出)馆址;2000年改作上海美术馆(2012年迁出)。2016年上海历史博物馆入驻原跑马总会大楼,经过保护修缮性改造,于2018年3月正式对外开放。

Shanghai History Museum

Shanghai History Museum is located at No.325 on West Nanjing Road. The building built in 1933 used to be Shanghai Race Club in the 1930s. From 1952, it had been jointly used by the Shanghai Museum and Shanghai Library. The two institutions were moved out in 1959 and 1997. In 2000, the Shanghai Art Museum was set here. It had operated at this location until 2012 Shanghai History Museum was moved to the former Shanghai Race Club Building in 2016. After the renovation, the museum officially opened to the public in March, 2018.

环球港双子塔：
工人新村的诞生
Global Harbor:
The Founding of New Villages

环球港凯悦酒店的四十七层高空室外露台，拥有360°无敌视野。曹杨新村在其西北，是上海在1949年后兴建的第一个工人新村，对我国居住区建设产生了深远影响。

The outdoor terrace is located on the 47th floor in Hyatt Regency's of Global Harbor. It has a superlative view of Caoyang Community in the northwest. It was the first estate built for wokers after 1949, and had a profound impact on the urban planning and residential construction.

建筑师	查普门·泰勒
建成时间	2008 年
建筑功能	酒店、办公、公寓、零售、展览
建筑高度	245 米
建筑层数	49 层
地址	中山北路 3300 号
开放时间	11:30—次日 01:00
公共交通	地铁 3、4、13 号线金沙江路站；公交 67 路金沙江路中山北路站，公交 69 路华东师大站

Architect	Chapman Taylor
Built	2008
Building features	Hotel, Office, Apartment, Retail, Exhibition
Total building height	245m
Total floors	49
Address	3300 Zhongshan Rd. (N)
Recommended viewing time	11:30-01:00 (next day)
Public Transportation	Metro Line 3/4/13 Jinshajiang Rd. Station; Bus 67 Jinshajiang Rd./Zhongshan Rd.(N) Station; Bus 69 East China Normal University Station

1949年后，在"为工人阶级服务"市政建设方针指导下，上海市政府首先选定中山北路以北、曹杨路以西，靠近沪西工业区的农田征地建房，即曹杨新村。1952年曹杨一村迎来首批1002户纺织、五金等行业的劳动模范和先进工作者入住，并成为国家首批外事接待单位。

曹杨新村从建成起就拥有超前的社区概念。沿环浜（初建时结合区域内原虬江及其支流开挖）扇形分布的曹杨一村，以现代"邻里单元"的理念，将至少10%的土地定为开放公共空间，每隔3栋楼设小花园，每个邻里配置步行范围内的社区服务，同步建设学校、影院、商场等公共设施，堪称"15分钟社区生活圈"。白墙、红顶、绿树的整体风格亦成为宣传画、年画讴歌美好生活的经典形象。1977年，曹杨9个新村基本建成。住区外围及东侧真如货运铁路支线沿线布置小型工业生产设施，在1998年铁路停运后用作农贸市场，2021年建成复合型高线公园"百禧公园"。

曹杨新村也在不断地改造、扩建、兴建。1962年一村由原2层住房加高至3层，1966年和1980年续建5层和6层住房，1989年新建10层商住楼。2009—2011年，进行过三次整改修缮；2019年底至2021年，70年历史的曹杨一村完成"成套化"改造，在保护历史建筑风貌和原有城市肌理的原则下，采取"一户一方案"，增强居民的参与感和认同感。如今的"幸福曹杨"，在保护传承历史文化风貌的同时，构建起多元化的邻里生活新模式，是上海市"十四五"规划的城市更新示范区之一。

After 1949, as the industrial development sped up, the population of manufacturing labor increased rapidly. Housing started to be a considerable issue. To bring convenience to workers' daily life, Shanghai Municipal People's Government decided to build a large cluster of apartment complexes nearby the Huxi (means west of Shanghai) Industrial District, located to the north of North Zhongshan Road and the west of Caoyang Road. This living cluster was named Caoyang Community. The Caoyang No.1 Community was completed in 1952, residing 1002 families of those who had earned medals for their great contribution and dedication in the areas of textiles and hardware.

Many advanced planning concepts and ideas assimilated into the design of fan-shaped Caoyang No.1 Community, as "Neighborhood Units", for an example, were adopted from Europe and America. More than 10% of the properties inside each community were featured with open space, landscapes, and convenient stores. Shopping malls, schools, and movie theaters were also nearby. It was the first residential model of "Community life within 15 Minutes." White walls and red roofs complemented by rows of green trees, often were the artists' favorite subject in their work at that time. The whole project with 9 villages were successfully completed in 1977. Along the Branch Line of Zhenru Freight Rail on the outskirts of the Caoyang Community, some small to medium sized the manufactories were put in functions. This industrial region was turned into a farmer's market after the railway transportation stopped in 1998. And then in 2021, this area was renovated into a highline park called "Centennial Park."

To meet the increasing population and to satisfy the residents' needs, it was necessary for Caoyang Community to be expanded and remodeled. Two-story Caoyang No.1 Community was raised to a three-story building in 1962. Such expanding continued in 1966, 1980, and 1989, from 3 stories to 5, 6, and 10. Between 2009 and 2011, the renovation took place for the third time. From the end of 2019 to 2021, the "One Set" renovation was completed in Caoyang No.1 Community while the traditions and overall styles lasted for 70 years had not been removed. Each residential unit in the complex had its own unique design featuring even newer facilities and more fashioned installations. The residents were engaged to participate and contribute their ideas in the renovation projects. Today's Caoyang Community, is widely known by its diverse living models and styles.

曹杨一村
Caoyang No.1 Community

环球港、内环线与地铁3、4号线
Global Harbor, Inner Ring Road and Metro Line 3&4

曹杨新村村史馆

位于花溪路 199 号,于 2013 年 8 月 26 日建成,为上海市爱国主义教育基地。村史馆共 5 层,每层约 600 平方米,集党史教育、村史教育、劳动教育、社会实践、教育服务等为一体。在这里可以充分了解曹杨新村的历史变迁,感受曹杨人的生活方式与劳模精神,展望未来数字化社区的发展。

Caoyang Community History Museum

The museum is located at No. 199 on Huaxi Road and was completed on August 26, 2013. The museum has 5 stories, and each floor is about 600 square meters. This place is the patriotic education base in Shanghai. The study halls of the history of Chinese Communist Party, constructional timeline of Caoyang Community, Model Workers Lecture Hall, Community Volunteer Service, and education information center are all included in one entity. Here you can fully understand the historical changes of Caoyang Community, feel the lifestyle and model spirit of Caoyang's people, and look forward to the development of a digitizing community is expected in the near future.

长宁龙之梦：
公共服务设施的蜕变
Cloud Nine:
The Construction of Public Service Facilities

万丽酒店的空中大堂和自助餐厅分别位于长宁龙之梦二十五层和二十六层。2005年建成的长宁龙之梦裙房为上海首个城市交通枢纽型商业综合体。

The Renaissance Hotel's grand lobby and buffet restaurant are located on the 25th and 26th floors of Cloud Nine. The design of Zhongshan Park and the urban structure around it are closely related to the planning of the urbanization. The Cloud Nine built in 2005 is the first commercial complex in the city transportation hub.

建筑师	ARQ 建筑事务所
建成时间	2008 年
建筑功能	酒店、办公、零售、公交车站
建筑高度	238 米
建筑层数	58 层
地址	长宁路 1018 号
开放时间	18:00—22:00
公共交通	地铁 2、3、4 号线中山公园站；公交 54、88 路中山公园站，公交 13、67 路中山公园地铁站

Architect	Arquitectonica
Built	2008
Building features	Hotel, Office, Retail, Bus Terminal
Total building height	238m
Total floors	58
Address	1018 Changning Rd.
Recommended viewing time	18:00-22:00
Public Transportation	Metro Line 2/3/4 Zhongshan Park Station; Bus 54/88 Zhongshan Park Station; Bus 13/67 Zhongshan Park Subway Station

长宁龙之梦与苏州河
Cloud Nine and the Suzhou Creek

历史上，中山公园附近就是铁路、公交集散地。1906年，工部局辟筑白利南路（现长宁路），1915年为连接上海北站（沪宁线）和南站（沪杭线）建造沪杭铁路沪西段，设梵皇渡车站（1916，1935年更名上海西站，1989年更名长宁站，1997年拆除，2000年建成轨道交通中山公园站）。

中山公园的前身是1860年英国人建造的兆丰花园，最初为私人庄园。1879年，极司菲尔路（现万航渡路）以北的兆丰花园土地出让给圣约翰书院，后名圣约翰大学。1914年，工部局在尚存的兆丰花园基础上兴建兆丰公园（也称极司菲尔公园、梵皇渡公园），成为上海近代著名的公园之一，1944年更名中山公园。1949年后，在保留原有自然景观的基础上，上海市政府多次对公园进行局部改建，使之成为大型游园和文化展览活动的重要场所之一。近年来为配合长宁路拓宽，地铁2号线、北横通道施工及绿化景观工程，公园陆续辟出开放空间，为周边区域提供了良好的生态环境。2022年，中山公园经过新一轮改造，拆除围墙，将万航渡路原先的英式园门和公园绿化融入市政道路，与此前开放的苏州河华政步道及华政校园连成一片，以公园文化、校园文化带动市民文化。

中山公园周边汇集了上海近代历史上诸多的文化教育资源。1952年，高等教育为适应蓬勃发展的社会主义建设需要，全国院系调整。圣约翰大学旧址上成立华东政法学院（合并复旦大学等9所院校的法律、政治和社会系，2007年更名华东政法大学）；圣玛利亚女校旧址上成立上海纺织工业学校（后改为上海纺织高等专科学校，1999年并入东华大学）。1951年，在大夏大学原址上成立华东师范大学（合并大夏大学文、理、教育学科与光华大学等相关科系），是我国创办的第一所师范大学。1983—2000年中山公园西面凯旋路一侧为上海大学美术学院（2000年迁往宝山校区）。

高校为中山公园及周边区域增添了公共服务设施和文化气息，便捷的公共交通改善了出行体验，城市公园协调人与自然的关系，使中山公园商圈成为兼具观赏游览、休憩娱乐、文化教育、颐养居住的城市副中心。

In the history, the area around the Zhongshan Park had been the center of railroads and other land transportations. In 1906, the SMC built the Brenan Road (now Changning Road). The segment on Shanghai-Hangzhou railway route connecting the Shanghai North (Shanghai-Nanjing Railway) and South (Shanghai-Hangzhou Railway) was completed in 1915, followed by the site of Jessfield Station in 1916. Jessfield Station was renamed as Shanghai West Station in 1935 and as Shanghai Changning Station in 1989. The station was removed in 1997. On its ground, metro station of Zhongshan Park was set up in 2000.

The former Jessfield Park, a private garden built by Hogg Brothers in 1860, is the predecessor of Zhongshan Park. In 1879, part of the property, to the north of Jessfield Road (now Wanhangdu Road) of the park was donated to the Episcopal Church of the United States, which was renamed as St. John's College later.

In 1914, the SMC built Zhaofeng Park on the remaining property of Jessfield Park. It was one of the most famous concession parks in modern Shanghai. In 1944, the park was renamed Zhongshan Park. After 1949, the municipal government reconstructed the park for several times, while the natural scenery in the park had been kept. With the expansion, this place eventually had become an oversized amusement park, and suitable for many kinds of cultural and art exhibitions. In recent years, cooperating with the widening of Changning Road and the new constructions of Metro Line 2 and Beiheng passageway, Zhongshan Park has successively been updated alongside these projects to provide pleasing views and a fine ecological environment. In 2022, the century-old Zhongshan Park once again had another round of renovation. The enclosure walls were torn down and the original English-style gate on Wanhangdu Road and some landscapes were set into the net of city transportation lines. Thus, Zhongshan Park, walking trails along the Suzhou Creek, and the campus of East China University of Political Science and Law were all connected as an entity. The projects of renovations and reconstructions over Zhongshan Park has shown the vast potential in continuous developments for better city parks and gardens.

Educational resources had been collected and stockpiled in the area around the Park. To improve the education in universities and colleges, the education department of Chinese government made a great revison and adjustment in 1952. East China College of Political Science and Law, later renamed as East China University of Political Science and Law in 2007, was established on the site of St. John's College. Nine departments of law studies, political studies, and social studies from the former St. John's College and Fudan University were merged into this new school. Shanghai Textile College, which was established on the site of St. Mary's Hall, merged into Donghua University in 1999.East China Normal University established in 1951 by combining the Great China University and Kwang Hua University on the site of Great China University is the first normal university founded after 1949. Shanghai Academy of Fine Arts (Shanghai University) had stood on Kaixuan Road to the west of Zhongshan Park since 1983, and then moved to Baoshan campus in 2000.

The establishments of colleges around Zhongshan Park have brought many public service facilities to this area, and then, emphasized the cultural and academic environments. All convenient transpotations and pleasant scenic views along with the many entertainments and shopping center have ameliorated the quality of residents' daily life.

长宁龙之梦全景
Panoramic view of Cloud Nine

长宁来福士广场

原址为美国圣公会创办的私立圣玛利亚女校（成立于1881年，1923年迁于此）。女校历史上曾出现过近代上海的杰出女性，如教育家俞庆棠，作家张爱玲等。1952—2006年，先后为上海纺织工业专科学校、上海纺织高等专科学校和东华大学纺织学院长宁校区。2009年遭强拆。2017年经由凯德地产保护性开发建成长宁来福士广场，原址修复了始建于1920年的钟楼礼拜堂，复刻重建思邠堂、格致楼、膳堂和思卜堂及大草坪等开放为城市公共空间。

Raffles City Changning

St. Mary's Hall was an American missionary school. It was founded in 1881 and moved to Brenan Road in 1923. Many outstanding women in modern times in Shanghai came from this school, such as Chintong Yui, an educator, and Eileen Chang, a novelist. From 1952 to 2006, Shanghai Academy of Textile Industry operated one after another on the site of St. Mary's Hall. In 2009, the site of St. Mary's Hall was forcibly demolished for some reason. Raffles City Changning was built by the protective development of Capitaland in 2017. The bell tower, built in 1920, remained on the site with other rebuilt buildings give the people an open green lawn for relaxing.

5F
开放空间
Open Space

芮欧百货
Réel Department Store

芮欧百货:
静安寺的前世今生
Réel Department Store:
A Memory of Jing'an Temple Area

芮欧百货五楼的露台可以看到大片公园绿地及古刹金顶。静安寺地区凭借其优越的地理位置,南接徐家汇、北上沪宁高速、西通大虹桥,东回"十里洋场"外滩起点,成为"国际静安"的核心商圈。

The outdoor terrace of Réel Department Store is located on the 5th floor. Jing'an area, neighboring with Xujiahui to its south, directly accessing the Shanghai-Nangjing Expressway from its north, connecting Hongqiao CBD on its west, and bounding the famous Bund on its east, is surely benefited from its advantageous location, has become a core of commerce. It is not only Shanghai's Jing'an, instead, it is "International Jing'an."

建筑师	株式会社久米设计
建成时间	2012 年
建筑功能	零售
建筑高度	24 米
建筑层数	5 层
地址	南京西路 1601 号
开放时间	10:00—22:00
公共交通	地铁 2、7、14 号线静安寺站,公交 15、20 路静安寺站,公交 71 路常德路站

Architect	Kume Sekkei
Built	2012
Building features	Retail
Total building height	24m
Total floors	5
Address	1601 Nanjing Rd. (W)
Recommended viewing time	10:00-22:00
Public Transportation	Metro Line 2/7/14 Jing'an Temple Station; Bus 15/20 Jing'an Temple Station; Bus 71 Changde Rd. Station

静安公园百年梧桐大道
Centennial Sycamore Avenue in Jing'an Park

静安寺地区历史上曾有"静安八景",即文人墨客传咏的赤乌碑、陈朝桧、虾子潭、讲经台、涌泉、绿云洞、芦子渡、沪渎垒,除涌泉外,余皆无存。昔日涌泉浜以静安寺为界,称东涌泉浜(现南京西路)和西涌泉浜(1952年于西涌泉浜旁填筑永源浜路,90年代更名永源路)。1862年起,英租界越界筑路,修建自泥城浜(现西藏中路)至静安寺的马道(静安寺路),静安寺周围形成一年一度四乡云集的静安寺庙会(1881年至1963年),沿途逐渐经营起申园(1882)、张园(1885)、愚园(1890)等娱乐性花园。静安寺东面的"愚园"(原址为1870年代的公一马房,1911年筑路因园命名愚园路)由于市中出现了更先进的综合性游乐场,以及城市化带来的沪西地价上涨,于1917年关闭,地产商收购其土地后建近代里弄、公寓府邸。保留至今的是原花园大门处的爱丁顿公寓(1936,现常德公寓)和地块旧改后平移修缮的愚园路81号(1947,现中共地下组织斗争史成列管暨刘长胜故居)。

1933年,静安寺地区出现了当时上海滩最摩登的休闲娱乐场所"百乐门",底层为店面,二层为舞池和宴会厅,建筑和设备均领时代之先,出现第一支华人专业爵士乐队,被誉为"远东第一乐府"。1954年改造为电影院。2017年,本着"修旧如旧"的修缮原则,进行结构加固和更新改造,还原其装饰艺术风格的建筑外观、内部U字形回马廊和弧形楼梯,重现弹簧木地板、玻璃舞池等舞厅设施。

静安公园原址是1898年工部局辟建的静安寺公墓。1953年,市政府将其改建成公园。园内古树众多,林荫蔽空,由32株百年悬铃木组成中央大道,东部草坪保留了原先公墓设施改建而成的大理石亭,20世纪八、九十年代增设尊师重教纪念碑和主题雕塑,西部则叠山理水、因势利导地营造出都市自然景观。公园将历史上的静安八景浓缩成园中园,以小见大,形成公园独具特色的休闲人文景观。1999年,公园配合地铁2号线建成静安寺下沉式商业广场,下沉广场的屋顶作为公园绿地的延伸,复合利用土地。2017年起,配合地铁14号线建设开启新一轮改造,期待成为更加立体、高效、生态以及可持续的城市公共空间。

一个多世纪以来,静安寺见证了上海城市化进程中的乡间马车、有轨电车(1908)、公共汽车(1922)、无轨电车(1928)、轨道交通(2000),从百年前的静安寺电车栈(1952年起上海市电车公司、电车一场、公交三分公司,2002年拆除,现静安嘉里中心北区东侧)到新时代的上海机场城市航站楼(2002)、静安寺公交枢纽站(2010),以及即将建成市中心最大的轨交枢纽站;从传统庙会,到大型商业综合体及国际消费中心城市示范区,让我们见证静安寺梦圆福地。

There used to 8 famous sceneries in the area of Jing'an Temple. Unfortunately, none of them exists today except the Bubbling Well. In old days, Yongquanbang (Bubbling Well) Creek was divided by Jing'an Temple into east and west sides. The east side of Bubbling Well is now West Nanjing Road. In 1952, Yongyanbang Road was filled and built over West Bubbling Well Creek, and in the 1990s, renamed as Yongyuan Road. The concession began to build roads outside of its territory from 1862. Bubbling Well Road was started from the Defence Creek (now Middle Xizang Road) and ended at Jing'an Temple as a horse path. Between 1881 and 1963, there was a large temple fair held in here once a year. The fair stimulated businesses along the road. Quite a few entertaining places came into this area, like Shen Yuen Garden (1882) Chang-su-ho Garden (1885), Yu Yuen Garden (1890). Yu Yuen Garden, to the east wing of Jing'an Temple, was located at the site of a stable in the 1870s. Later in 1911, Yuyuan Road was built and named after the garden. Yu Yuen Garden had to close its business in 1917, because of the even larger and more attractive entertainments in the city and quick rise of the land appreciation in the west side of the city. Some real estate companies purchased its property and developed this area into different residence. Of which, there are two places of historical interests have been kept till nowadays. One is Eddington House (now Changde Apartment), built in 1936; the other one is the former Residence of Liu Changsheng built in 1947 on Yuyuan Road.

In 1933, the most modern entertainment place Paramount Ballroom was launched in the Jing'an Temple area. It had shopfronts on the ground floor, ballroom & banqueting hall on the second floor, and a bridge room in the tower. Architecture and building service were ahead of the times. It is famous for the first professional Chinese jazz band and was well known as "No. 1 music hall in the Far East." In 1954, the Paramount Ballroom was transformed into a movie theater. In 2017, after the structural reinforcement and lavish renovation, the Paramount Ballroom reappear its glistening Art Deco details, the U-shaped corridor & the curved stars, the spring floor and the crystal floor provide a luxury and comfort experience.

The original site of Jing'an Park is the Jing'an Temple Cemetery built by the SMC in 1898. In 1953, it was changed to the park. In the park, the central avenue is composed of 32 century-old sycamore trees. The eastern lawn retains the marble pavilion transformed from the original cemetery facilities. In the 1980s and 1990s, the monument and sculpture of respecting teachers and valuing education were added. In the west of park, rockeries and waters forms a natural landscape of central city. The park contains a small garden with the historical "Jing'an Eight Scenes," which provide a unique leisure and cultural landscape. In 1999, a sunken square was built at the northwest corner of Jing'an Park, cooperated with the project of Metro Line 2. The roof of the sunken square serves as an extension of the green land. Since 2017, in cooperation with the construction of Metro Line 14, the area was being on new renovation with a more efficient, ecological and sustainable urban public space to be expected.

For more than a century, the Jing'an Temple has witnessed the rural carriage, tramcars (1908), autobus (1922), trolley bus (1928), and rail transit (2000) in the process of urbanization in Shanghai. The site of former Shanghai Tramways was pulled down in 2002, now for the Jing An Kerry Centre. There are also Shanghai Airport City Terminal (2002) and Bus Hub (2010) around Jing'an Temple, and the largest rail transit hub of central Shanghai is under construction. From traditional temple fairs to modern commercial complexes and international Jing'an, let's bless the dream of city Jing'an coming into reality.

静安寺与延安高架路
Jing'an Temple and Yan'an Elevated Road

芮欧百货与静安寺
Réel Department Store and Jing'an Temple

芮欧百货全景
Panoramic view of the Réel Department Store

- 大胜胡同 Dasheng Hutong
- 复旦大学附属华山医院 Huashan Hospital, Fudan University
- 上海龙之梦大酒店 Longemont Hotel Shanghai
- 上海国际贵都大饭店 Hotel Equatorial Shanghai
- 长宁龙之梦万丽酒店 Cloud Nine Shopping Mall & Renaissance Shanghai Zhongshan Park Hotel Center
- 会德丰国际广场 Wheelock Square
- 环贸港 Global Harbor
- 百乐门 Paramount
- 静安寺 Jing'an Temple
- 真如城市副中心 Zhenru Subcenter
- 久光百货（久百城市广场） Joinbuy City Plaza of Shanghai
- 铜仁小区 Tongren Community
- 静安嘉里中心 Jing An Kerry Centre
- 上海商城 Shanghai Center
- 恒隆广场 Plaza 66
- 静安寺广场 Jing'an Temple Square
- 静安公园 Jing'an Park

静安寺

据南宋《云间志》载，静安寺创建于三国赤乌年间，1216 年自吴淞江畔迁至法华镇芦浦沸井浜畔（现址），屡记屡修。寺内"应天涌泉"（沸井），从 19 世纪末开始，因多次道路修筑，号致地下水源被切断，于 1960 年代加盖填埋。90 年代末地铁施工时挖掘出围栏、雕塑等文物，现存于上海历史博物馆。1998—2021 年底静安寺扩建完工，东西两侧分布钟、鼓楼，24 米高的梵幢和 63 米高的佛塔坐落于寺院的东南和西北角，钟楼底层为重新挖掘的涌泉，名"潜龙古水"。

Jing'an Temple

One of the local chronicles in the Nan Song Dynasty stated that the Jing'an Temple was built in 247 AD and moved to the bank of Bubbling Well, which is the current location in 1216. There had been many reconstructions over the time. Around the end of 19th century, the bubbling well lost its source of underground water. Finally, in 1960, this well was filled with dirt and capped. In the 1990s, some of archaeological artifacts, like well fences and statues, were dug out during the constructions of underground subway. Now these artifacts are kept in Shanghai History Museum. From 1998 to 2021, Jing'an Temple had been undergoing another round of renovation and expanding. The temple with new appearance is featured by clock tower on the east side and drum tower on its west. A 24-meter tall Brahma building sits on the southeast of the court while a 63-meter pagoda stands on the northwest.

41F
开放空间
Open Space

新锦江大酒店
Jin Jiang Tower Hotel

新锦江大酒店：高架道路的建设
Jin Jiang Tower Hotel: The Construction of Elevated Roads

位于酒店 41 层的蓝天旋转餐厅，可以 360°俯瞰魔都街市，透过环绕餐厅的落地玻璃幕墙，延安高架路和南北高架路蜿蜒如城市动脉贯穿东西南北。

The public space of the new Jin Jiang Hotel is located on the 41st floors. There is a revolving restaurant with a 360-degree view and an administrative lounge with an outdoor terrace. In the evening, visitors can see the lights on Huaihai Road flashing in the distance and watch vehicles running on Yan'an Elevated Road and North-South Elevated Road.

建筑师	王董建筑师事务所
建成时间	1988 年
建筑功能	酒店
建筑高度	153 米
建筑层数	43 层
地址	长乐路 161 号
开放时间	11:30—22:00
公共交通	地铁 13 号线淮海中路站；公交 24、41 路瑞金一路淮海中路站，公交 26 路长乐路陕西南路站
Architect	Wong Tung Group of Companies (WT)
Built	1988
Building features	Hotel
Total building height	153m
Total floors	43
Address	161 Changle Rd.
Recommended viewing time	11:30-22:00
Public Transportation	Metro Line 13 Middle Huaihai Rd. Station; Bus 24/41 Ruijin Rd. (No.1)/Huaihai Rd.(M) Station; Bus 26 Changle Rd./Shaanxi Rd.(S) Station

新锦江大酒店与延中绿地广场公园
Jin Jiang Tower Hotel and Yanzhong Square Park

新锦江大酒店与锦江饭店
Jin Jiang Tower Hotel and Jin jiang Hotel

上海开埠后，开始建造现代城市道路。此后的大部分时间里，公共租界、法租界与华界，各自为政，互不统辖，形成"一市三治"的市政格局。加之上海是江南水乡，河渠纵横，池塘密布，沿河筑路和填浜筑路，致使出现许多歪歪扭扭、弯弯曲曲的马路，对街道风貌产生深远的影响，也使全市交通呈现"局部有序，全局无序"的面貌。

1949年后，政府对肇嘉浜进行改造使其成为市区重要的交通干道。50年代后期，上海开始编制《上海市高速干道系统规划》。1959—1960年，市区外围的中山北路、中山西路、中山南路连接拓宽为中山环路，形成通向周边城镇和近郊工业区的道路起点。

80年代初，上海开始启动"三横三纵"主干道的修建工程，成为90年代中心城区"申"字形高架路的基本骨架。1983年为配合第五届全运会拓宽四平路，2000年拓宽华山路，历时十多年的主干道工程全部竣工。

随着城市交通压力逐步增大，1985年市规划局提出在中山环路上建造高架机动车专用道，从地面向空中发展。1990年，上海市建委批准内环线高架道路规划方案。1994年内环高架路全线建成通车，成为"中国高架第一环"，通过南浦大桥和杨浦大桥把浦东、浦西的交通连为一体。

由于历史成因，上海南北向干道少于东西向干道。1993年市政府决定建造纵贯南北的成都路高架（后名南北高架），沿线经过市中心人口最为稠密的南市、黄浦、静安、闸北，开启了上海城建史上空前规模的10万市民大动迁。1995年建成通车，两端分别接内环高架路。南北高架六车道的设计，充分考虑到避让沿线诸多文物保护单位和历史建筑，如中国社会主义青年团中央机关旧址、中国劳动组合书记部和中共二大会址等多处革命旧址及妇女用品商店（培文公寓）、复兴公园等。

1995年，穿越中心城区的延安路高架采用分段施工、分步通车的方法建设。1996年建成西段（内环以西），1997年建成东段（石门一路以东），1999建成中段（石门一路至内环）。其中东段与中段的地面道路原本是由百多年前的洋泾浜（外滩至西藏中路）、长浜（金陵西路成都路至华山路）和柴兴浜（华山路至武夷路）填筑而成，延安路高架如龙舞一般穿越在高楼林立之间，由西向东经过虹桥地区、静安公园、上海展览中心、延中绿地、上海大世界、中汇大厦等历史建筑和城市景观，顺势拐至延安路高架外滩下匝道，弧形江景一览无余，曾被誉为"亚洲第一弯"。2008年因配合外滩通道综合改造工程，拆除另建外滩隧道，现下匝口正对外滩信号塔。

高架路的建设，不仅减轻了上海的交通压力，同时也加快了城市生活节奏。内环申字形结构对上海城市空间形态的发展具有决定性的意义，此后陆续延伸建成外环（2003）、中环（2015）等快速路网及郊环（绕城高速公路），成为上海的城市动脉。

After the open-up of port in Shanghai, many roads in the city were constructed. In most of the time, the common concession, French concession, and Chinese territories were independent from each other. This municipal structure was called "One City with Three Governances." The early main geographical feature in Shanghai was full of rivers and creeks. Almost all the roads were constructed along the water banks or built by filling and covering the beds of waters. The city's transportation network appeared in malformation.

After 1949, the municipal government decided to reform the Zhaojiabang Creek and make it a main road in the city's transportation net. In later 1950s, Planning of arterial road in Shanghai get on to the agenda. Between 1959 and 1960, North Zhongshan Road, West Zhongshan Road and South Zhongshan Road on the outskirts of the city were widened and connected. This effort brought all three roads together as a loop throughout to nearby suburbs.

In the beginning of the 1980s, the engineering project called "Three Horizontals and Three Verticals" was carried out. This project focused on main transportations roads in Shanghai. The "申"-shaped transportation frame appearing in the 1990s was the result from the project. The Three Verticals refer to three routes in the direction of north to south. The east route is Siping Road-Wusong Road-East Zhongshan 1(2) Road-South Zhongshan Road, the middle route is Gonghexin Road-North Chengdu Road-Middle & South Chongqing Road-Luban Road, and the west route is Caoyang Road-North Jiangsu Road-Huashan Road-North Caoxi Road. The Three Horizontals refer to three lines going in the direction of west to east. The north line is Changning Road-Changshou Road-Middle Tianmu Road-Haining Road-Zhoujiazui Road, the middle line is West Yan'an Road-Middle Yan'an Road-East Yan'an Road, and the south line is Hongqiao Road-Zhaojiabang Road-Xujiahui Road-Lujiabang Road. To better serve the purpose of the 5th National Sports Event, Siping Road was widened in 1983. Then in 2000, the widened Huashan Road was put in use. The engineering project over the city's transportation main roads was finished.

As the speedy growing of the nation's economy and the ameliorating of people's living qualities caused heavier and busier traffics, the government suggested to build elevated tracks only for motor vehicles on the whole Zhongshan Road pushing urban development towards in the air and sufficiently utilizing the space was a wise idea. In 1990, the plan of constructing Inner Ring Elevated Road was approved. In 1994, the construction was completed. This was the very first elevated transportation structure in the whole nation. The east side and the west side of the Huangpu River are connected by Nanpu Bridge and Yangpu Bridge.

For some reasons from the old days, there were many fewer routes going in the direction of north to south. In 1993, the municipal government decided to build elevated Chengdu Road (now North-South Elevated Road) on this direction. Since this route goes through highly populated districts like Nanshi, Huangpu, Jing'an, and Zhabei, more than a hundred thousand residents in these areas had to be relocated. This is the first massive relocation in the history of urban development in Shanghai. In 1995, the elevated Chengdu Road was finished. This route with 6 lanes connects the city's inner loop on both ends. Fortunately, many historical buildings and relics along the route were not touched and kept well.

In 1995, the project of building elevated Yan'an Road started. To cause less inconvenience of people's daily life, this project was divided into several segments. In 1996, the west segment was finished. In 1997, the east part, to the east of Shimen 1st Road was put in use. The entire route started to function as the middle segment completed in 1999. The ground-level roads of the east and the middle segments were originally built over the Yangjingbang Creek (from the Bund to Middle Xizang Road), Changbang Creek (from West Jinling Road/Chengdu Road to Huashan Road), and Chaixingbang Creek (from Huashan Road to Wuyi Road). The Yan'an Elevated Road, like a dancing dragon, unrestrainedly passes through some historical buildings and urban sceneries like Hongqiao Area, Jing'an Park, Shanghai Exhibition Center, Yanzhong Square Park, Dashijie, the former Chung Wai Bank Building from west to east. The curve is astonishingly smooth as it reaches the exit ramp near the Bund where every scenic point of view can be reached. In 2008, this curved ramp was abolished in cooperation with the comprehensive reconstruction on the Bund.

The elevated transportation road system not only has eased the stress of the traffic in the city, but also has paced up the urban life. The 申-shaped frame of transportation network inside the inner loop has contributed the significant concepts towards the continuous improvement of utilizing the space in the city of Shanghai. The characteristics of urban life in Shanghai indicate that developments of the city has entered in a new era followed by the completion of Outer Ring Expressway (2003), the Middle Ring Road (2015) and Shanghai City Highway.

延安高架路
Yan'an Elevated Road

上海音乐厅

位于延安东路523号，原名南京大戏院，建于1930年，是近代中国建筑师范文照设计的新古典主义风格建筑，也是中国第一座音乐厅。1950年更名为北京电影院，1959年更名为上海音乐厅。此后数十年，一直是上海音乐活动的中心，多次举办上海之春音乐节、国际广播音乐节、星期广播音乐会等。2003年配合延安高架路的拓宽，上海音乐厅整体抬高、平移和修缮，2019—2021再次修缮，是上海最具历史性的专业音乐演出场所与海派文化地标。

Shanghai Concert Hall

Located at No. 523 on East Yan'an Road, the Shanghai Concert Hall, formerly known as the Nanjing Grand Theater, was built in 1930. It is a neoclassical-style building designed by modern Chinese architect Robert Fan. It was the first concert hall in China, renamed as Beijing Cinema in 1950 and Shanghai Concert Hall in 1959. It has been the center of all musical events in Shanghai, such as Shanghai Spring International Music Festal (since 1959) and Shanghai Weekly Radio Concert (since 1982). On April 15, 2003, in order to cooperate with the widening construction of Yan'an Elevated Road, the SCH whole building was lifted up, and then was shifted 66.46 meters southward. It was undergoing another round of the renovation between 2019 and 2021. Now, sitting on the new site, the SCH is certainly another landmark, with historical legacy and Shanghai Syle.

24F
开放空间
Open Space

斯格威铂尔曼大酒店
Pullman Shanghai Skyway Hotel

斯格威铂尔曼酒店：打浦桥地区旧城更新
Pullman Shanghai Skyway Hotel: The Urban Regeneration of Dapuqiao Area

酒店的行政酒廊位于 24 层，开阔的视野向南可远眺浦江，向北可见打浦桥地区更新后的城市繁华景象。

Located on the 24th floor, the administrative lounge of the Hotel has the view of the Huangpu River to the south and urban prosperity of Dapuqiao area to the north.

建筑师	B+H 建筑事务所
建成时间	2007 年
建筑功能	酒店
建筑高度	226 米
建筑层数	52 层
地址	打浦路 15 号
开放时间	11:30—22:00
公共交通	地铁 9 号线打浦桥站；公交 17 路打浦路斜土路站，公交 43 路打浦桥站
Architect	B+H Architects
Built	2007
Building features	Hotel
Total building height	226m
Total floors	52
Address	15 Dapu Rd.
Recommended viewing time	11:30-22:00
Public Transportation	Metro Line 9 Dapuqiao Station; Bus 17 Dapu Rd./Xietu Rd. Station; Bus 43 Dapuqiao Station

打浦桥地区
Dapuqiao area

旧区改造是城市发展的重要课题，在斯格威铂尔曼酒店的上空可以直观感受城市更新对南北高架路沿线打浦桥地区的改变。

打浦桥，本为跨肇嘉浜的一座桥，位于今徐家汇路、瑞金二路口，1946年填没日晖港以东部分河道时拆除，后作为该区域地名。肇嘉浜东接黄浦江，西连蒲汇塘，是上海老城厢正中的干流，也是向松江府运粮的内河。进入近代，由于大兴土木造成肇嘉浜上游阻塞，人口增长，倾倒大量垃圾使其失去航运功能。抗战爆发后，大批难民聚集于此形成水上棚户，肇嘉浜成了名副其实的臭水浜。

1949年后，上海市政府开始大力整治打浦桥地区。1954—1956年，在日晖港建造大型泵站抽排污水，填没肇嘉浜，建成通往徐家汇的林荫大道——肇嘉浜路，大道中心成为群众休息游乐的街心花园，同时建设商业网点，初步建成区域内商业中心。1957年开始拆除打浦桥地区的危棚简屋，营建新住宅，但显然跟不上棚户区人口的增长速度，甚至棚户区面积呈扩大趋势。改革开放为旧城改造带来新机遇，1978年建造的14层打浦大楼是卢湾区最早的棚户改造项目，也是当时区内最早的高层住宅。90年代后，该地区发生翻天覆地的变化。1992年通过土地批租形式推动旧改，1994年建成的海华花园小区是改革开放以来上海第一个通过吸引外资实现旧区改造的项目，称"海上第一块"，成为上海城建史上的重大突破。1997年底，打浦桥地区最后一间危棚简屋铲倒，其间完成徐家汇路拓宽（1993）、日晖港综合治理工程（1995）、打浦路拓宽（1996）、瑞金二路（1996，原为日晖东路，填埋日晖港后扩建），陆续建成卢湾体育馆（1996）、金玉兰广场（2004）、日月光中心广场（2010）等，在建造高档商务楼和住宅小区的同时，开启"石库门文创园"之先。

田子坊，前身是1930年建造的志成坊，老式石库门里弄。泰康路是打浦桥地区的一条小街，在1980年代起的环境治理和90年代工厂企业转制改革中，道路两侧大批厂房闲置。打浦桥街道率先提出，利用废弃厂房招商营建泰康路工艺品特色街的设想，通过街道、居民、艺术家的共同努力，自下而上地以租赁、转让、置换等方式，把旧厂房、旧民宅改建成艺术家工作室。低廉的租金和便利的交通条件，吸引了一批以画家、摄影家为代表的文化人入驻，画家黄永玉为其起名"田子坊"，遂成为艺术与生活、传统与现代融合的典范。当前，同质化的商铺不断稀释田子坊的魅力，蜂拥的游客带来经济效益，也使房租上涨，环境愈发嘈杂，居民和艺术家流失，城市空间的生命力如何延续，值得各方探讨。

The regeneration of the existing historical areas is an important subject and task in urban development. Viewing over the Pullman Shanghai Skyway Hotel, we can clearly tell how the urban structural renewals have brought the significant impact onto the Dapuqiao area along the North-South Elevated Road. Dapu Bridge, once crossed the Zhaojiabang Creek, was in the intersection of today's Xujiahui Road and Ruijin 2nd Road. The bridge was torn down when some segment of the creek was drained and the bed was filled in 1946; however, the bridge's name, Dapu, was kept for the area around the creek. Zhaojiabang Creek was a major flow in Shanghai's old Town. Its east end connects the Huangpu River and the west end linked the Puhui Pond. In old days, it was an important water path on which commissary was delivered to the Songjiang prefecture. With the continuous increasing of the population and constructions, more and more construction rubbish and municipal waste had been dumped into the creek. The upstream of the creek was blocked up, and eventually the creek lost its function of conveyance. Many refugees settled down in this area after the War of Resistance against Japanese Aggression broke out. The area had become an overcrowded shanty town. Zhaojiabang Creek was nothing but a foul canal.

After 1949, the municipal government began to vigorously regulate Dapu Bridge area. From 1954 to 1956, the polluted water was pumped and drained out, and the creek bed was filled. A beautiful boulevard leads to Xujiahui, Zhaojiabang Road, was built. There was a garden full of shades and flowers in the middle section of the road. Lots of businesses started to set up along the road. The commerce center inside this area preliminarily took the shape. The shabby shelter homes began to be removed in 1957, but the new residential buildings were still far from the demands from the increasing population. The shantytown even appeared to be enlarged. Fortunately, the Reform and Opening-up promoted the vast residential construction in this area. The 14-floor Dapu Building was built in 1978. It was the tallest dwelling structure ever built in this area. Dapuqiao area have had enormous changes since 1990. In 1992, the ordinance of property leasing was passed. Haihua Garden Community, completed in 1994, was thevery first constructional project funded by coporation overseas. This project was the breakthrough in Shanghai urban construction history. Finally in 1997, the last shelter home in Dapu area was eradicated. During the same period, the widening of Xujiahui Road (1993) and Dapu Road (1996), the comprehensive governance in Rihui Port (1995), the construction of Luwan Sport Stadium (1996), the Golden Magnolia Plaza (2004), and SML Center (2010) were completed. Meanwhile, the changing of Shikumen was presented to people while high-class business buildings and apartment complexes had been built up unceasingly.

Tianzifang, formerly known as Zhichengfang, built in 1930, is an old-fashioned Shikumen lane. Taikang Road once was a small street in the Dapuqiao area. The environmental governance in the 1980s and the restructurings of manufacturing in the 1990s left many factories space unoccupied and unused. The management agency of Dapuqiao area proposed a plan of utilizing those empty spaces. The various leasing methods, low rents, and the convenient location attracted many artists to set private studios. "Tianzifang" (田子坊) was the handwritten inscription by the famous painter, Huang Yongyu, "Tianzifang" has set an example of combing the traditions and the fashions and integrating the arts with realities. The economic benefits brought in by the influx of tourists caused increases of the property rents. Not only the environment has become clamorous, but also many shops and studios with the close styles and the similar types have been put up. The artists and nearby residents have already started to move away from here. Is such a loss bearable? How to generate the vitality of urban public facilities? We all must put in some serious thoughts and considerations.

田子坊鸟瞰
Aerial view of Tianzifang

斯格威铂尔曼酒店与卢浦大桥
Pullman Shanghai Skyway Hotel and Lupu Bridge

陈逸飞工作室

位于泰康路210弄2号,原是1930年代日本人建的机械零件工厂。1998年前后,陈逸飞、尔冬强、吉承等艺术家、设计师将几处旧工厂改造为工作室,并作为会客厅。自此,这片落寞的工业厂房开始迎来艺术中心、画廊、工作室的艺术创意聚会。虽然画廊及工作室如今已经搬迁,但旧址免费向公众开放,仍然是面向全球旅人的会客厅,为人们展现上海老城工业建筑的再生利用。

Chen Yifei Studio

Located at No. 2, Lane 210, on Taikang Road, the studio was originally a mechanical parts factory built by the Japanese in the 1930s. Around 1998, artists and designers like Chen Yifei, Er Dongqiang, and Ji Cheng transformed several old factory buildings into art studios and reception halls. This desolate industrial plant was replaced by galleries and studios and became the heart of creative art. Although the gallery and studio have been relocated, the original site is still open to the public. It welcomes tourists from everywhere and presents the evidence and advantage of reviving the deserted constructions in the old regions in Shanghai.

第 3 章
垂直生长时期的城市文明新貌

1990 年代浦东开发开放以来，上海城市建设进入史无前例的快速发展时期，城市面貌日新月异，建筑高度不断攀升。从城市制高点浦东陆家嘴、苏州河沿岸、外滩、后世博园区、北外滩和虹桥商务区都可以感受到上海的蓬勃朝气。

以陆家嘴中央商务区为代表的浦东新区，创造了震惊世界的城市建设奇迹，建设量之大、建设速度之快突破了人类以往对城市发展规律的认知。如此进程对上海的都会风貌核心区产生了三个突出结果：

一是城市迅速拔高。1980 年代开始，上海高层建筑的数量迅猛增长，至 2001 年底高层建筑的总面积已跃居世界第一。今日上海在数量、平均高度和最高建筑方面，都是全球当之无愧的"摩天都市"。

二是城市现代化转型快。在基础设施建设方面，城市道路、桥梁、轨道交通以及其他市政设施飞速迈入现代化，在改变城市风貌的同时也极大丰富了市民的都市生活。

三是城市空间拼贴感强。虽然核心区早在近代就形成较为成熟的城市功能分区和结构，但早年租界各自为政的建设方式导致城市缺乏统一的规划和协调。1990 年代以来的快速建设难免忽略引导和管控，对区域结构和城市风貌产生了极大冲击。

这些因素共同导致上海核心区的拼贴感，一方面创造了城市空间的丰富性，另一方面也产生了较为明显的割裂现象，城市风貌特征在空中视野下显得尤为突出。

摩天楼是世界现代建筑史的奇迹，也是见证一个城市发展史的标志，不断刷新的高度代表着城市生长的力量。凭借铁路和水运便利通达的优势，近代上海从一个传统江南水乡市镇发展成为中国东南沿海经济贸易枢纽和国际化大都市。如今上海的"摩天都市"特征彻底形成，其经济地位也不断攀升，摩天大楼已经成为这座世界级城市的重要名片，推动上海在新时代背景下的国际金融中心建设，助力中国与世界经济共振。

CHAPTER 3
The New Look of Urban Civilization in the Period of Vertical Growth

After the opening of Pudong Area in the 1990s, the pace of the urban constructions in Shanghai accelerated. The appearance and structure of the city has kept changing continuously. The average height of high-rise buildings set the new record time and again. There have been full of Shanghai spirit of enthusiasm and energy from the city top of Lujiazui, the Suzhou Creek riverside, the Bund, the Post-Expo Area, and the Hongqiao CBD.

Especially in Pudong, centered by Lujiazui CBD, the speed and number of new constructions are beyond imagination. The developments of Pudong have influenced the core area of Shanghai city.

The first is the rapid increase in buildings' heights. The area of high-rise buildings in Shanghai has been on top of the list in the whole world in 2001. Shanghai is known as "Skyscraper Metropolis" in terms of the number of and the height of the buildings.

The city's modernization is the second impact. Infrastructure constructions like roads, bridges, rail transit and other municipal facilities have presented advanced techniques and technologies and new fashions. The urban life has been greatly enriched.

The third is the strong feeling of urban space "collage." Although the core area of the city had already formed a relatively mature functional zone and urban structure as early as modern times, the independent regulation and management from the concessions had caused the city lack of unified planning and coordination. Also, because of the inadequacy of guidance and control, the hasty constructions in the 1990s resulted both regional and the entire city's structures disorderly.

All these factors together led to a sense of "collage" in the core area. While abundant urban space was created, the appearance of the city from the aerial view seemed unruled and fragmented.

The skyscraper is a miracle in the world history of architecture. It is a key witness of a city's development. Continuous record breaking in buildings' heights shows a city's progressive potential. Shanghai is no longer a small traditional town on the water. Relying on advantages in railway and water transportations, Shanghai has developed into an international metropolis and a hub of commercial trades on the east coast of China. The skyscrapers have become an important landmark of this global city, promoting Shanghai as an international financial center in the new era and helping China resonate with the world economy.

东方明珠和上海中心：
陆家嘴的高度攀升
Oriental Pearl Radio
& TV Tower and
Shanghai Tower:
The Rising of
Lujiazui CBD

263M
开放空间
Open Space

东方明珠
Oriental Pearl Radio & TV Tower

东方明珠
Oriental Pearl Radio & TV Tower

主观光层位于上球体263米标高,其上层267米标高是可以容纳350位游客用餐的旋转餐厅,其下层259米标高的悬空玻璃观光走廊,是全球唯一的360°全透明观光廊;位于顶球351米标高的太空舱,可以鸟瞰整个城市美景。

The main observation deck occupies three floors inside the upper large sphere at 263 meters, 267meters (revolving restaurant) and 259 meters (transparent glass corridor). The highest observation deck named "Space Capsule" is inside the top sphere at 351 meters, and it affords an overall view of the city.

建筑师	华东建筑设计研究院有限公司
建成时间	1995年
建筑功能	电视塔、观光
建筑高度	468米
建筑层数	16层
地址	世纪大道1号
开放时间	09:00—21:00(每日最晚售票时间为20:00)
公共交通	地铁2、14号线陆家嘴站;公交81、85陆家嘴地铁站,公交82路陆家嘴环路丰和路站
Architect	ECADI
Built	1995
Building features	TV Tower, Sightseeing
Total building height	468m
Total floors	16
Address	1 Century Ave.
Recommended viewing time	09:00-21:00 (cut-off time for ticket is 20:00 daily)
Public Transportation	Metro Line 2/14 Lujiazui Station; Bus 81/85 Lujiazui Subway Station; Bus 82 Lujiazui Ring Rd./Fenghe Rd. Station

东方明珠和上海中心 | Oriental Pearl Radio & TV Tower and Shanghai Tower

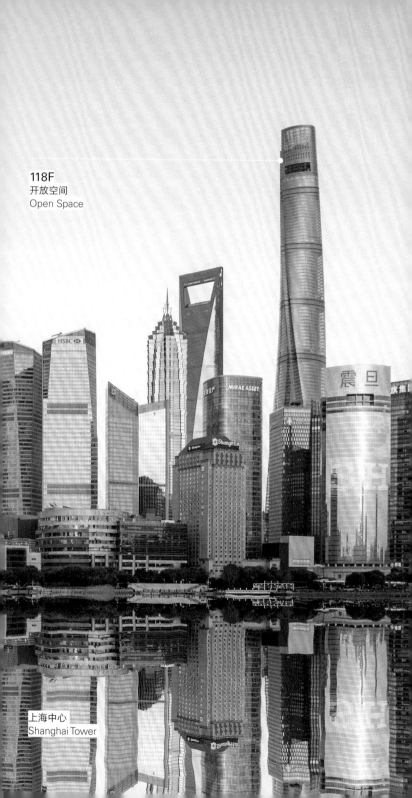

上海中心
Shanghai Tower

"上海之巅"观光厅位于上海中心118层，垂直高度达546米，是目前上海向公众开放的最高公共空间。从这里俯瞰上海，可以全方位感受"天空之城"的魅力。

The observation deck of Shanghai Tower, named Top of Shanghai, is located on the 118th floor, with the height of 561 meters. It is the highest public space open to civilians. Standing in front of the glassy wall, you will be amazed and fascinated by this enchanted "City in the Sky."

建筑师	晋思建筑设计事务所 同济大学建筑设计研究院（集团）有限公司
建成时间	2016 年
建筑功能	酒店、办公、会展、观光、零售
建筑高度	632 米
建筑层数	119 层
地址	世纪大道 100 号
开放时间	08:30—22:00（每日最晚售票时间为 21:30）
公共交通	地铁 2 号线、14 号线陆家嘴站；公交 583、939 花园石桥路东泰路站，公交 791、870 路陆家嘴环路东泰路站
Architect	Gensler, TJAD
Built	2016
Building features	Hotel, Office, Exhibition and Sightseeing, Retail
Total building height	632m
Total floors	119
Address	100 Century Ave.
Recommended viewing time	08:30-22:00 (cut-off time for ticket is 21:30 daily)
Public Transportation	Metro Line 2/14 Lujiazui Station; Bus 583/939 Huayuan Shiqiao Rd./Dongtai Rd. Station; Bus 791/870 Lujiazui Ring Rd./Dongtai Rd. Station

陆家嘴与浦西
Lujiazui and Puxi

地理上，陆家嘴是黄浦江与苏州河（吴淞江）相汇后形成的嘴状冲击沙滩；历史上，明代文学家陆深的旧居以及陆氏的祖茔在此，故得名。上海开埠后，陆家嘴与浦西租界仅一江之隔，从老地图上的船厂路、烟厂路、烂泥渡路、草塘弄等地名可见此地较多码头、工厂与工棚简屋。沿江一带主要为轮渡码头（陆家嘴至延安东路）和江南造船厂（1954，前身为1914年招商局船厂和1862年英商祥生船厂，2005年整体搬迁，改建为"船厂1862"时尚休闲空间），可供市民游玩的公共设施仅为浦东公园（现东方明珠电视塔和滨江大道一部分，原址为1859年外国水手公墓及早年英联船厂和日商纱厂旧址）。1991年9月1日，东方明珠电视塔打下第一根桩机，1994年建成陆家嘴地区最早的地标建筑。1997年建成陆家嘴中心绿地，2000年世纪大道建成通车，形成超高层建筑群。其中被称为"上海三件套"的建筑分别是：1999年竣工的468米金茂大厦、2008年竣工的492米环球金融中心、2016年竣工的632米上海中心。

上海中心是上海的第一高楼，步入地下一层的上海之巅展示厅，动画浓缩了自1840年上海开埠以来180余年外滩滨江、浦东陆家嘴天际线的发展。登上位于118层的观光厅回望百年外滩，感受浦东开发与浦西城市建设联动。浦东陆家嘴正与隔江对望的虹口北外滩、黄浦百年外滩同心聚力、共筑上海"黄金三角"核心商务区战略高地。

陆家嘴地区陆续建成海洋水族馆（1999）、正大广场（2002）、上海国金中心（2010）、船厂1862（2018）、浦东美术馆（2021）等休闲文化场所。对比市内的其他地区，在超高层摩天大楼的视觉冲击下，陆家嘴尚缺乏多样性与能持续吸引市民关注的公共活动。近年来通过不断优化地下通道、建设空中环形天桥等立体人行步道系统，改善区内的步行环境及楼宇可达性。希望未来的规划和建设，可以通过植入更多小尺度、混合型的复合空间，形成更加宜人的环境，更好地激发城市活力，使陆家嘴成为集萃万象的多元化生活场所。

In geography, Lujiazui is the alluvial shoal formed by the Huangpu River and Suzhou Creek. Lujiazui was named after Lu Shen, an intellectual in the Ming Dynasty. The former home of Lu Shen was in here, so was his ancestral graveyard. Lujiazui was apart from the concessions by the Huangpu River. After the opening of Shanghai port, many corporations from overseas had built wharves and factories in Lujiazui. Lots of shacks for labors' residing had also appeared. The road names on some old maps can reflect such a state: Chuanchang (shipyard) Road, Yanchang (tobacco factory) Road, Lannidu (a muddy ferry) Road, Caotang (grasses & ponds) Lane, etc. Before the developmental plan of Pudong was carried out, Lujiazui was mainly a ferry terminal just nearby the CSSC (1954, formed by an amalgamation of 1914 China Merchants Shipyard and 1862 Boyd & Co., moved to Changxing Island in 2005, the former site was renovated as the Harbor City "MIFA 1862"), and Pudong Park, sitting on the old site of the Foreign Sailors' Cemetery, Shanghai Dockyards, and a Japanese textile mill, was the only public facility where residents could enjoy some relaxing time. On September 1st, 1991, the first pile of the Oriental Pearl Radio & TV Tower was put in the ground. After 3 years, the tower was completed, and it has become an eye-catching landmark in Lujiazui. The Lujiazui Central Greenbelt was built in 1997. Century Avenue started to function in 2000. The pattern of high-rise buildings around the Central Greenbelt and the traffic loop has formed. The reputated "Three Skyscarpers" refers to 468-meter-tall Jin Mao Tower (1999), 492-meter-tall Shanghai World Financial Center (2008), and 632-meter-tall Shanghai Tower (2016).

Shanghai Tower is the tallest building in the city. The Top of Shanghai exhibition hall began from the basement level 1. The animation presentation provides the tourists the history of developments in Lujiazui and the Bund over more than 180 years. The observation deck is on the 118th floor. Looking out to the Bund, we can percept how the developments in Pudong had impacted on urban restructurings and renewals in Puxi. The unified economic growth in the area of Yangtze River Delta and Yangtze River Economic Zone also has been encouraged by the great progress in both Pudong and Puxi. Achievements in Lujiazui in Pudong and the North Bund in Hongkou, the Bund in Huangpu have prompted the "Golden Triangle" into a core of commerce and has driven Shanghai onto the track of international economy and finance.

While Lujiazui has become the center of the businesses and finance, lots of office buildings, residential zones, restaurants, and other service and amusement facilities have turned up. Places like Shanghai Ocean Aquarium (1999), Super Brand Mall (2002), IFC (2010), MIFA 1862 (2018), Museum of Art Pudong (2021) are very popular to the local residents and tourists from everywhere. However, with the comparison with other regions in the city, Lujiazui needs to work on the varieties and endurance in attracting people to be active. Some efforts, like optimizing the traffic routes underground, building elevated pedestrian walkway, constructing convenient paths to different destinations, and improving the surrounding atmosphere, have been made in recent years. Smaller-scaled and multi-functional spaces are expected in the future development plan. Making Lujiazui a diverse and enjoyable living environment for all people is a never-ending goal.

陈桂春住宅

又称"颍川小筑",位于陆家嘴中心绿地南侧。建于1917年,为中西合璧风格的庭院式民居建筑。1992年浦东开发之初,经社会各界人士奔走呼吁,得以保留;后经修缮保护与功能置换,1996年作为陈家嘴开发陈列馆对外开放;2010年,上海吴昌硕纪念馆迁址于此。

Chen Guichun Residence

Chen Guichun Residence is located in the south of Lujiazui Central Greenbelt. The construction was completed in 1917. It is traditional courtyard residence combined with Chinese and Western styles. Quite early in the development of Pudong in 1992, scholars and citizens appealed for the preservation of this building. In 1996, it was renovated for Lujiazui Development Exhibition Hall. In 2010, Shanghai Wu Changshuo Memorial Hall was moved to Chen Guichun's Residence.

浦东开发陈列馆

位于浦东大道 141 号，始建于 1956 年，原为东昌工人俱乐部，浦东文化馆。1990 年至 2000 年，为上海市人民政府浦东开发办公室、浦东新区人民政府。2009 年配合市政道路建设整体北移 9.13 米、东移 2 米。2010 年 4 月 18 日，浦东开发 20 周年之际，正式开放为浦东开发纪念馆。"141" 的谐音也寓意了浦东开发"一是一、二是二、实事求是"所传递的诚信精神。

Pudong Development Exhibition Hall

Loacated on 141 Pudong Avenue, this building was built on 1956, formerly Dongchang Workers' Club and Pudong Cultural Center. From 1990 to 2000, it was used as Pudong Development Office of Shanghai Municipal People's Government and the People's Government of Pudong New Area. In 2009, the building was moved 9.13 meters to the north and 2 meters to the east due to the municipal road construction. On April 18th, 2010, the 20th anniversary of the implementation of the Pudong development, Pudong Development Exhibition Hall opened to the public. Something quite interesting needs to be mentioned. The number of this building, 141, is the homophone and connotation of "One is one, two is two. Never stop seeking truths from facts"—the honest spirit throughout the entire long course of the Pudong development.

宝格丽酒店：苏州河水岸治理
Bulgari Hotel Shanghai: The Comprehensive Environmental Treatment of the Suzhou Creek

宝格丽酒店位于上海总商会（1915）旧址旁，在四十七层酒吧露台向外望，天际线下，苏州河蜿蜒绵亘，自西向东在外白渡桥汇入黄浦江，外滩源、陆家嘴尽收眼底。

IL Bar is located on the 47th floor of Bulgari building, beside the former site of Shanghai General Chamber of Commerce (1915). The Suzhou Creek flows from west to east, and join the Huangpu River at the Waibaidu Bridge, with a panoramic view of the Rock Bund and the Lujiazui CBD.

建筑师	安东尼欧·西特利奥、派翠西亚·维尔
建成时间	2017 年
建筑功能	酒店、公寓
建筑高度	150 米
建筑层数	48 层
地址	河南北路 33 号
开放时间	17:00—次日 01:00
公共交通	地铁 10、12 号线天潼路站；公交 14、19、25 路天潼路河南北路站

Architect	Antonio Citterio, Patricia Viel
Built	2017
Building features	Hotel, Apartment
Total building height	150m
Total floors	48
Address	33 North Henan Road
Recommended viewing time	17:00-01:00 (next day)
Public Transportation	Metro Line 10/12 Tiantong Rd. Station; Bus 14/19/25 Tiantong Rd./Henan Rd.(N) Station

上海邮政博物馆与河滨大楼
Shanghai Postal Museum and Embankment Building

相比宽阔壮丽黄浦江，苏州河的尺度更加温婉宜人，30座建于不同年代的桥梁与九曲十八弯的苏河光影互相交叠，勾勒出一幅两岸生活和历史记忆的长卷。

苏州河，是吴淞江流入上海市区内河段的别称，也是江浙航运转向黄浦江和长江入海口的要道。1840年代起，在外滩靠近苏州河的区域，先后有二十几个国家建立过领馆。19世纪末苏州河沿岸地区被划入扩张后的公共租界范围，两岸陆续出现近代城市公用事业。1864年大英自来火房在苏州河以南、泥城浜以东建煤气厂（现西藏路桥东侧，1953年上海市煤气公司，2001年建高层办公楼），1864年成立的公济医院1877年迁至乍浦路桥北堍（1953年上海市第一人民医院，1996年迁往武进路，拆除老建筑后2018年建成苏宁宝丽嘉酒店），1880年英商上海自来水公司在江西路桥（1942年拆除）修建输水管（1921年桥南堍建自来水公司办公楼，现为上海市供水行业协会和城投水务有限公司），上海成为我国最早享受工业文明带来便捷的近代城市。

在公用事业呈河流导向建设的同时，由于当时内河交通的便捷性和比铁路运输更低的成本，进入20世纪之后，苏州河沿岸由东向西绵延成以四行仓库（1921，1932，1990年代用作商业办公，2014年建成四行仓库抗战纪念馆）为代表的内河港口仓储区，以面粉厂、火柴厂为代表的早期民族资本加工企业和纺织工业高度密集的沪西工业区。它们与沿岸的商会、俱乐部、私宅、学堂、教堂、清真寺等形成丰富而多元化的水岸历史积淀。其中，宜昌路救火会（1932，现普陀消防支队）曾是沪西地区的制高点；四川路桥南堍的第一加油站（1948）是中国第一家国营加油站；四川路桥北堍的上海邮政总局（1924，现上海市邮政局暨上海邮政博物馆）拥有"远东第一大厅"的邮政营业厅。

也正是因为水路的四通八达，自1930年代起，上海周边坐船而来的难民，在苏州河两岸俗称"三湾一弄"（朱家湾、潭子湾、潘家湾、药水弄）的贫民窟安身立命，逐渐形成日后中心城区北部最大的棚户区。人口的集聚和城市的勃兴，生活污水和工业废水不断地排入苏州河，致使河底淤泥堵塞，水体局部出现黑臭，至1970年代，苏州河几乎全线黑臭浑浊，外白渡桥外侧的苏州河与黄浦江交汇处甚至出现了一条泾渭分明的黑（苏州河）黄（黄浦江）分界线。

1980年代，上海"决心把苏州河治理好"。1988年开始的合流污水治理工程成为苏州河整治的重要前奏。1996年全面启动苏州河环境综合整治，集水质、水利、防汛、环保、环卫、绿化等内容为一体。2001年，苏州河上首次举办龙舟赛，拉开了一年一度上海苏州河城市龙舟赛的序幕。1999—2006年，全面建成中远两湾城，昔日棚户区变身内环线内规模大的现代化生态居住区。近年来,随着"一江一河"战略规划的推进,水岸联动的图景越来越生动。至 2020 年末,苏州河中心城区 42 公里岸线基本贯通开放,河水变得清澈,鱼虾重现河底,形成两岸宜人的自然生态和滨河人文新景观。2022 年底,苏州河游轮正式启航,使广大市民有了更多亲近"母亲河"的机会。

Comparing to broad and magnificent Huangpu River, Suzhou Creek's flow is more soothing and amiable. Thirty bridges built in different years set of the lights and shadows in the ripples. Life and history along the shores have precipitated in long-flow of the Suzhou Creek.

Suzhou Creek is just one segment of the Wusong River as it flows in urban area in Shanghai. It is a main channel for Jiangzhe Shipping turning into the bayou of the Huangpu River and the Yangtze River. More than 20 countries had set up embassies or consulates on the Bund near the Suzhou Creek since 1840. At the end of the 19th century, the area around the banks of Suzhou Creek was incorporated in the expanded concessions. Some modern-styled urban public facilities started to show up along the shores. A gas plant was built in 1864 by Shanghai Gas Co. Ltd. at south of the Suzhou Creek and east of the Nichengbang Creek (now east side of Xizang Road Bridge, 1953 Shanghai Gas Company, 2001 rebuilt a tall office building). The General Hospital, which was assembled in 1864, was relocated to nowadays north side to Zhapu Road Bridge in 1877. In 1996, the hospital was moved to Wujin Road. After the old building was torn down, Bellagio by MGM Shanghai was built on the site in 2018. Shanghai Waterworks Company funded by English began to install water pipes on Jiangxi Road Bridge (demolished in 1942). The office building of Shanghai Waterworks Company was built at south side of the bridge in 1921. Today, Shanghai SMI Water Co. Ltd. sits on this location. Shanghai was the first city in modern history benefitted from the industrial civilization.

Because of the convenience and lower cost of the transports by inland rivers, the area nearby the shores of Suzhou Creek had become a dense industrial district since the beginning of 20th century. Flour mills, match factories, processing plants, and textile enterprises from the early-age national capitals had emerged. This region centered by the Joint Trust Warehouse which later was used as a business management office in the 1990s and reconstructed as the Memorial. Combined with industrial developments, chambers of commerce, clubs, schools, churches, private residences, and mosques had formed modern life with rich culture and diverse styles. The fire station built in 1932 on Yichang Road was once the tallest building in Huxi area. The gas station, set in 1948, on south side of Sichuan Road Bridge was the first state-run business in gasoline. Shanghai General Post

Office, built in 1924, on the north side of Sichuan Road Bridge, was reputed by its business lobby – "The First Lobby in Far East."

With the easy transportation on the water, many refugees from nearby towns and countryside started to settle down on the banks of Suzhou Creek since the 1930s. The zone had eventually been formed into the largest slum to the north of the urban. Continuously increasing population and growing industries had caused sewages and discharges pouring into the Suzhou Creek. The water was severely polluted and the watercourse was clogged by silt on the bottom. By the end of the 1970s, water in the Suzhou Creek totally turned into dark color with unpleasant foulness. There was a clear dividing line between two waters when the Suzhou Creek and the Huangpu River meet.

In the 1980s, Shanghai Municipal started a plan to regulate the Suzhou Creek. The first step of the regulation project was confluence treatment for polluted water in 1988. The comprehensive environmental remediation was launched in 1996. Tasks of optimizing water quality, water conservancy, flood control, ecological balance, environmental-friendly improvements, and landscaping were all involved. The first dragon boat race held in 2001 on the Suzhou Creek was the prelude of the annual national water sports in Chinese traditions. The project of Zhongyuan Liangwan Cheng Community was implemented and completed during 1999 and 2006. This project successfully and thoroughly removed shacks in the slums and developed that area into a largest ecological residence with modern fashions within the city's inner loop. In recent years, advanced by the strategy blueprints, "One River and One Creek," the sceneries along the shores of Suzhou Creek appears lively. By the end of 2020, the 42-kilometer-long shoreline was open to the public. Fish swim in the clear water, while people stroll in the fresh and crispy air. Suzhou Creek cruises were launched at the end of 2022. While enjoying the fun on the cruises, tourists embrace the heart of the "Mother River" and cherish the moments of joyful life.

遥望陆家嘴
Towards the Lujiazui

宝格丽酒店与苏州河
Bulgari Hotel Shanghai and the Suzhou Creek

M50 创意园

原为上海春明粗纺厂,是目前苏州河畔保留最为完整的民族工业建筑遗存之一,共拥有50余幢工业建筑。2000年起,逐步引进以视觉艺术和创意设计为主体的艺术家工作室、文化艺术机构和设计企业,成为上海具有标志意义的创意园区之一。M50西侧,原阜丰机器面粉厂和福新面粉厂旧址上建成商业综合体"天安千树"(2021,一期商场),保留并新老钟塔为电梯艺术塔,在开放式露台上栽种绿植,成为苏州河畔的新地标。

M50 Creative Park

Originally known as Shanghai Chunming Woolen Mill, M50 Creative Park is the most complete relics of national industrial buildings on the banks of the Suzhou Creek. Since 2000, artists' studios, cultural & art institutions, and designing enterprises, with visual art and creative design as the main emphases, have been gradually introduced, becoming one of the landmark creative parks in Shanghai. To the west of M50, built on the former site of Fou Foong Mill Co. "1000 Trees" has become a new landmark on the banks of Suzhou Creek.

外滩茂悦酒店：城市滨水空间
Hyatt on the Bund: Open Space of the Huangpu Waterfont

位于酒店三十二楼和三十三楼的酒吧和露台，是观赏外滩历史风貌和浦东高楼大厦的绝佳位置，从而感受滨水空间的多元化和城市空间的连续性。

The Vue Bar and terrace at Hyatt on the Bund is the perfect place to enjoy the views on both sides of the Huangpu River where the diversity of waterfront space and the continuity of urban space came into sight.

建筑师	HOK 建筑事务所
建成时间	2007 年
建筑功能	酒店
建筑高度	约 120m
建筑层数	33 层
地址	黄浦路 199 号
开放时间	11:30—次日 01:00
公共交通	地铁 10、12 号线天潼路站；公交 37 路大名路闵行路站，公交 22 路闵行路长治路站
Architect	Hellmuth Obata Kassabaum (HOK)
Built	2007
Building features	Hotel
Total building height	120m
Total floors	33
Address	199 Huangpu Road
Recommended viewing time	11:30-01:00 (next day)
Public Transportation	Metro Line 10/12 Tiantong Rd. Station; Bus 37, Daming Rd./Minhang Rd. Station; Bus 22 Minhang Rd./Changzhi Rd. Station

外滩茂悦酒店与世界会客厅
Hyatt on the Bund and the Grand Halls

外滩本是黄浦江边的滩涂，后成为近代上海第一个租界。英租界江岸南起洋泾浜（现延安东路）北至苏州河处的李家厂（亦作"李家庄"，今北京东路外滩一带）。因老城厢小东门一带的江滩为"里滩"，这段便得名"外滩"。外滩沿江建筑始建于1849年，按照土地章程，沿江建筑和涨潮线之间必须留下滩地作为迁道，建浮码头通过栈桥与岸线相连，就此形成外滩地区观光堤岸的基本走向。

早期黄浦江堤岸为木桩篱笆，1860堤岸朝江中拓展，江畔种植树木，填滩建造公园（1868，上海最早的公共公园，1936年名外滩公园，1945年名春申公园，1946年名黄浦公园），至1886年，堤岸内侧扩建人行道，道旁设座椅供行人休息。外滩林荫道东端的黄浦公园门口从1950年代起设有公交终点站（圆形广场），多条线路通往西区的徐家汇、新华路、天山新村、以及东北区的五角场、提篮桥、杨树浦路等。

1950年以前，外滩堤岸一直都没有防汛墙，台风大雨江水倒灌是常态。1959年在原驳岸压顶建砖砌防汛墙，1963年建成钢筋混凝土L形防汛墙，此后的外滩改造都与防汛墙和道路改造紧密相连，防汛墙和观光平台也成为外滩的重要景观。80年代，由于市区住房紧张和休闲场所匮乏，防汛墙成了年轻人的约会场所，"情人墙"之名源出于此。1989—1993年，外滩防汛墙岸线向江心外移，建成钢筋混凝土双层空箱式结构的新防汛墙，厢体下部为外滩停车库，上部为外滩观光平台，全长1.8公里。与同期实施的外滩建筑群景观照明改造，日夜相交辉映。改造拆除黄浦公园大门、围墙，开放成为以人民英雄纪念塔（1994）为主体的外滩观光带起点。

2000年外滩开启一体化工程，连接南外滩和北外滩，形成连续的滨江岸线。2008年起，配合世博会的交通改造，先后拆除延安路高架"亚洲第一弯"、吴淞路闸桥（1991）、延安东路天桥（1982，1993年曾部分拆除），2010年建成外滩地下通道，开放更多地面空间与景观。

外滩江岸分设水文站（1912）和气象站（1884，1957年停用）。1908年建成49.8米高的外滩气象信号塔，于1993年整体东移20米，修缮后内设外滩历史陈列室，成为见证外滩百年气象的地标。外滩历史上，出现过多座纪念碑和铜像、雕塑，至20世纪中叶几乎无存，仅有延安东路附近的纪念碑基座（1924，欧战纪念碑暨胜利女神像，1941年被日军拆毁）1960年被移除。1993年国庆前，南京东路外滩矗立起陈毅市长的塑像，2010年命名"陈毅广场"，是外滩滨江的中心地带。

轮渡航线是外滩滨水空间的基础设施。1911年开始的东铜线，1935年撤销，码头处的巴夏礼铜像1943年拆除。1935年北京路外滩设双层浮船轮渡码头，下层为浮桥出入口，上层为水上饭店，是当时市区段最完善的轮渡码头之一。每逢盛夏，外滩到吴淞口的夜班轮渡成为最时髦的消夏方式，在 60 年代停办，1979 年又恢复"浦江游览"，推动了黄浦江水上旅游业务的兴起。90 年代之前，轮渡几乎是市中心往返浦东和浦西唯一的交通工具，"陆延线"（1956，延安东路至陆家嘴轮渡码头，1968 年和 1976 年两次扩建，1997 年配合延安路高架建设南移至金陵路口）高峰时段的繁忙拥挤达到不可承受之重。

外滩滨水空间从小舢板和小火轮开始，因水而岸，贯穿上海近现代城市发展的重要时间节点和空间进程，不仅能看到浦江两岸的风貌缩影，更能寻找出历史传承和城市基因，是承载记忆的城市核心滨水区。

The Bund was a piece of shoal along the Huangpu River. The first concession was set in here in modern Shanghai. The English concession started its southern border at Yangjingbang Creek (now East Yan'an Road) and stretched all the way to Lijiachang (now East Beijing Road nearbyYuanmingyuan Road) next to the Suzhou Creek as its northern line. Since the segment of the riverbank close to Xiaodongmen in the Old Town was called "Litan" (the Inner Bund), this part then was called "Waitan" (the Outer Bund). The constructions on the Bund began in 1849. According to the land regulations, all constructions must be spaced from the high tide line. Floating piers were connected to the shores by trestle bridges. This structure became the basis of later observation embankment.

In the early age, the embankment along the Huangpu River was wooden fence. In 1860, the embankment was extended towards the river. Trees were planted on the banks, and then The Public Park was built in 1868 over the site of filled shoal. This is the earliest public park in Shanghai. The name of the park had been changed for several times: Waitan Park in 1936, Chunshen Park in 1945, and Huangpu Park in 1946. In 1886, pedestrian sidewalk was paved at the inner side of the embankment. Chairs for people taking rest were placed next to the sidewalk. In 1950, the bus station was put in at the entrance to the Huangpu Park. Many transportation routes from this point had led to different locations, like Xujiahui, Xinhua Road and Tianshan Communities in the west side of the city, as well as Wujiaochang, Tilanqiao, and Yangshupu Road in the northeast.

There weren't flooding control walls before 1950. In hurricane seasons, heavy rains often caused river water intrusion. A brick flooding control wall was built in 1959, and an L-shaped reinforced-concrete wall was constructed in 1963. Thereafter, the flooding control wall is one of the priorities in all reconstructions on the Bund. Just like the observation platform, the flooding control wall is also a pleasant scenery on the Bund. In the 1980s, residential space and public leisure facilities were so limited that many young dating couples had to choose this location for having their romantic times. The flooding control wall was nicknamed "the Lovers' Wall." Between 1989 and

1993, the flooding control wall was moved outwards, and redesigned as a double-tiered and reinforced-concrete hollow chamber. The bottom space of the flooding wall was used as a parking garage on the Bund while the top space was turned into an observation platform for visitors. The wall was 18 kilometers long. At the same time, a few of other projects were underway. Buildings on the Bund began to show off magnificent lighting effects. The Huangpu Park, with the removal of its entrance gate and the wall around the park, had reformed as the starting point of the touring journey on the Bund. Shanghai People's Heroes Memoria (1994) placed in the park is the hot spot for visitors.

In 2000, the integration project was enforced. This project restructured the Bund by joining the South Bund and the North Bund and forming an incessant shoreline. To prepare for the World Expo, some traffic structures had been altered since 2008. "The First Curve" on Yan'an overpass, Wusong Road gate bridge (1991), and East Yan'an Reoad Skyway (1982, 1993) were taken out. The Bund Tunnel was completed in 2010.

There were hydrological and meteorological stations on the shore of Huangpu River. The hydrological station was built in 1912. The meteorological station was built in 1884 and its service ended in 1957. The 49.8-meter-tall Gutzlaff Signal Tower, built in 1908, was shifted 20 meters eastward in 1993. The tower, a landmark that witnesses the achievements of meteorology studies over hundred years, was remodeled and featured with an exhibition hall presenting the history of the Bund. There used to be many monuments, statues, and busts on the Bund. However, they all were wiped out by the middle of 20th century except the base of the Cenotaphe (the Allied War Memorial, European War Victory Memorial, Angel or Goddess of Peace Memorial, Victory Angel Memorial) nearby East Yan'an Road. It was damaged by Japanese military in 1941, and was removed in 1960. Before the Independence Day in 1993, the statue of the former Shanghai mayor, Chen Yi (1901-1972), was set up on the Bund close by East Nanjing Road. That location was named "Chen Yi Square" in 2010, and has become the heartland on the Bund.

Ferry routes are infrastructure along the Huangpu River. The first ferry route, Dongtong Line is started in 1911, abandoned in 1935. The copper statue of Sir Harry Smith Parkes (1828-1885) on the pier of on Nanjing Road was removed in 1943. This was the first government-run ferry on the Huangpu River. In 1935, a two-tiered floating ferry pier was put up. The bottom stratum was the entrance and exit, and the top was a restaurant on the water. This pier was one of the most consummate ferry piers in the city at that time. Ferrying at the night from the Bund to Wusongkou became the fashionable joyride in summer. The joyful ferry was cancelled in 1960s. The resume of "Pujiang Cruise" in 1979 boasted the travel businesses on the Huangpu River. Before 1990, ferry was almost the only transportation method between Pudong and Puxi. Lujiazui-East Yan'an Road Line, a land route, was hardly bearable at the traffic peak even though it was widened twice in 1968 and 1976.

The developments of waterfront on the Bund started from sampans and steamboats. As time passed by, the presence, structure, and establishments have been ameliorated and enhanced significantly. Accomplishment on the Bund is one of the most important chapters in the history of urban development in Shanghai.

闵行路看外滩茂悦酒店
From Minhang Road to Hyatt on the Bund

外滩茂悦酒店 | Hyatt on the Bund

外滩茂悦酒店全景
Panoramic view of the Hyatt on the Bund

外滩源

位于黄浦江和苏州河交汇处,得名自 2002 年立项的"洛克·外滩",2010 年公共绿地和圆明园路特色景观(街基本建成开放。东起圆明园路和外滩 33 号花园,西至虎丘路(初名博物馆路),这 11 栋历史建筑承载了重要的城市档案。其中,亚洲文会大楼(1932,现外滩美术馆)旧址是近代中国最早的博物馆(1874);兰心大楼(1930)旧址上的兰心剧院(1866)是上海最早的西式剧场;哈密大楼(1927,沙弥大楼)曾是《文汇报》报社;1946 年《新民报》晚刊(《新民晚报》前身)在虎丘路创立;1949 年上海人民广播电台入驻格林邮船大楼(1922)。2009 年上海半岛酒店开业(原址为上海友谊商店),成为外滩源新地标。

ROCK BUND

The ROCK BUND is located at the intersection where the Huangpu River and the Suzhou Creek, opened to the public in 2010. 11 buidlings from Yuanmingyuan Road and #33 on the Bund to Huqiu Road (original Museum Road) archived the city's history, of which included the first modern museum in China and the earliest modern theatre in Shanghai. It was also one of the origin places of the newspapering and broadcasting in Shanghai. In 1990, 22-story Wenhui Mansion was completed on the old site of Beth Aharon Synagogue (1927) on Huqiu Road. Before it was torn down in 2006, the building stood tallest in the area. In 2009, the Peninsula Shanghai was opened as a new landmark hotel on the former site of Shanghai Friendship Store.

上海当代艺术博物馆：后世博园区的发展
Power Station of Art: The Development of the Post-World Expo Park

上海当代艺术博物馆的眺江大平台位于五层，可以俯瞰整个世博园区、欣赏南浦大桥到卢浦大桥的沿岸风光，观看工业建筑群的前世与今生。

The platform on the 5th floor of the Power Station of Art provides a panoramic view over the former World Expo Zone, riverside sceneries along Nanpu Bridge and Lupu Bridge, and the industrial complex.

建筑师	同济大学建筑设计研究院（集团）有限公司原作工作室
建成时间	1987 年
改造时间	2012 年
建筑功能	艺术展览
建筑高度	50 米
建筑层数	7 层
地址	苗江路 678 号
开放时间	周二—周日 11:00—19:00
公共交通	地铁 4、8 号线西藏南路站；公交 55、89 路南浦大桥站，公交 18、45 路中山南路南车站路站
Architect	Original Design Studio, TJAD
Built	1987
Transformation	2012
Building features	Art Exhibition
Total building height	50m
Total floors	7
Address	678 Miaojiang Rd.
Recommended viewing time	Tuesday-Sunday 11:00-19:00
Public Transportation	Metro Line 4/8 South Xizang Rd. Station; Bus 55/89 Nanpu Bridge Station; Bus 18/45 Zhongshan Rd.(S) Station

上海当代艺术博物馆前身为南市发电厂（1897，原南市电灯厂，为上海第一家华人兴建的电厂，2007年搬迁），2010年上海世博会期间改造为"城市未来馆"。发电厂165米高的烟囱改成一只巨大的温度计，内部为独立的展览空间。

PSA所在区域位于上海县城之南。1867年，李鸿章创办的江南机器制造总局（1865，为近代中国第一厂）从虹口迁址沪南地区黄浦江边并开始建造船坞，拉开中国近代船舶工业发展的帷幕。辛亥革命后，称江南船坞为江南造船所，1952年更名为江南造船厂。2008年因世博会建设和自身发展需要，江南造船厂整体搬迁到长兴岛。同期置换改造的还有南市自来水厂的两个厂区，西部厂区（1902，华商创办的上海内地自来水厂，1956年合并附近原法商自来水厂后更名南市水厂）置换合并进东部厂区（1895，法商自来水厂）相邻地块，协调规划，让出黄浦江岸线景观通道和半淞园路城市绿地，百年老厂成为一座新型现代化水厂。这里记载了中国民族工业的早期印记，见证了上海从工业时代到信息时代的变迁。

作为2010年上海世博会城市最佳实践区，会后把低碳环保理念从单体建筑扩展至整个街区，形成以PSA为代表的文化创意产业园区。同时，滨江景观带向西延伸至徐汇滨江，沿岸建成世界乒联博物馆和中国乒乓球博物馆（2018），徐汇滨江规划展示中心、龙美术馆（2014，原1929年北票煤码头）、余德耀美术馆（2014，原龙华机场大机库）、油罐艺术中心（2019，原龙华机场航空储油罐）等公共文化设施。

隔江相望的世博会浦东片区，一轴四馆成为后世博上海新地标：世博源超广域型综合购物中心（世博轴）、中华艺术宫（原中国馆）、梅赛德斯奔驰艺术中心（原世博演艺中心）、世博会议中心（原世博中心）、世博展览馆（原主题馆）。在黄浦江东岸与世博大道之间，打造白莲泾公园（2010）、前滩世博休闲公园（2015）、后滩世博文化公园（2022，原上钢三厂），在建中的48米"双子山"将成为卢浦大桥边的新地标。

世博会推进了中心城区轨道交通网络建设，加速了浦江两岸再开发进程和城市更新脚步，优化了中心城区功能配置和空间结构，充分演绎了"城市 让生活更美好"的主题。

The site where the Power Station of Arts sits on nowadays used to be Nanshi Power Plant. In 1897, the Plant was built here. The factory was Shanghai's the first power plant entity built up by Chinese, and it was turned into the EXPO "Pavilion of Future." The 165-meter-tall chimney in the factory was furnished into a giant thermometer. Inside the space of the chimney was an exhibition hall. This landmark in the area is hardly to be missed.

The Power Station of Arts is in the south of Shanghai County. Kiangnan Arsenal, founded by Li Hongzhang in 1865, was the first manufacturer in modern China. In 1867, Kiangnan Arsenal was moved to the bank of Huangpu River from Hongkou. The construction of docks indicated that China had entered an era of shipbuilding enterprise. After Xinhai Revolution, the dock was named Kiangnan Shipyard, and it was renamed Jiangnan Shipyard in 1952. In 2008, to cooperate with the constructions of the World Expo and meanwhile to expand the production scale, Jiangnan Shipyard was relocated to Changxing Island. At the same time, two branches of Nanshi Waterworks had experienced reform. The western branch used to be a Merchant-Running waterworks since 1902, merged with the eastern branch, the original waterworks funded by French in 1895. The replacement land was then renovated into a scenic passage along the shore of Huangpu River and a green belt on Bansongyuan Road. The new waterworks with over a hundred year's history stands tall with modern technologies and scientific management in present days. It recorded the developments of Chinese national industries and witnessed Shanghai in transition from the period of industrial age to the internet time.

As the Urban Best Practice Area during the Shanghai World Expo in 2010, it continues to keep the concepts of low-carbon environment. From one single building to a wide stretch, an enterprise of cultural creation centered by PSA was formed. The waterfront scenic line has been extended to Xujiahui. Cultural facilities like the International Table Tennis Federation Museum (2018), China Table Tennis Museum (2018), Long Museum (2014, 1929 Coal Terminal), Yuz Museum (2014, the former hangar of Longhua Airport), Oil Tube Art Center (2019, the former aviation oil storage tank of Longhua Airport) have been built up and open to the public.

Across the Huangpu River, in the Pudong section of the World Expo, a "Four Halls and One Axis" pattern has been formed and become a post-Expo landmark in Shanghai. This special zone consists of the River Mall, China Art Museum, Mercedes-Benz Arena, Shanghai EXPO Center, Shanghai EXPO Exhibition and ConventionCenter. In 2022, Shanghai Expo Culture Park was built on the former site of Shanghai No.3 Steel Works. The 48-meter-tall "twin hills", which will become China's first artificial mountains, will also be key attractions in the southeast of the park at the former site of the World Expo 2010 along the Huangpu River.

The World Expo has advanced rail networks in the city, accelerated the reconstructions and renovations on both shores of Huangpu River, and enhanced the structures and configurations of urban public facilities. "Better City, Better Life" is no longer just a slogan, but a reality for people to experience.

上海当代艺术博物馆与卢浦大桥
Power Station of Art and Lupu Bridge

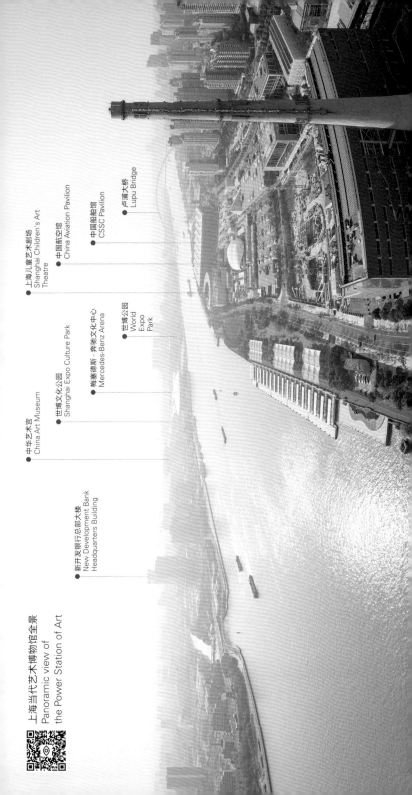

上海当代艺术博物馆全景
Panoramic view of the Power Station of Art

中华艺术宫

前身为建于2010年世博会中国馆,为世博轴(现世博源综合购物中心)中心地带的标志性建筑,以现代建筑形式与传统斗栱元素相结合,层层出挑、叠加,披誉为"东方之冠"。2012年,上海美术馆从人民广场南京西路迁至此地,更名为"中华艺术宫",成为上海新的城市文化名片。

China Art Museum

The China Art Museum used to be the China Pavilion during 2010 Shanghai World Expo. It is an iconic building on the center of the Expo. Being reputed as "the Crown of the East," the building is the composition of modern architectural vogue and traditional element of bucket arches. In 2012, Shanghai Art Museum was moved here from the People's Square and renamed China Art Museum.

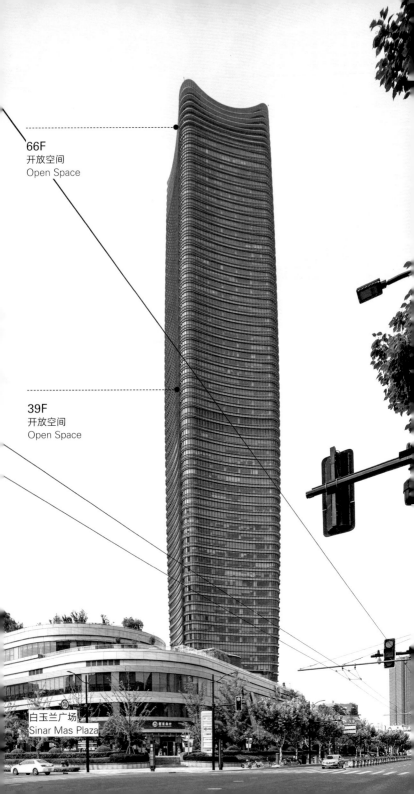

66F
开放空间
Open Space

39F
开放空间
Open Space

白玉兰广场
Sinar Mas Plaza

白玉兰广场：
北外滩的新生
Sinar Mas Plaza:
The Rebirth of
the North Bund

三十九层展厅和六十六层空中观景平台，呈现上海港的演化历史，可环视兼具历史底蕴与时尚魅力的北外滩和杨浦滨江城市空间。

The exhibition hall on the 39th floor and observation platform on the 66th floor provide a panoramic view over Shanghai Port with its abundance of history, the North Bund full of charms of fashion, and public facility spaces around the Yangpu waterfront.

建筑师	SOM 建筑设计事务所、华东建筑设计研究院有限公司
建成时间	2017 年
建筑功能	办公、酒店、零售
建筑高度	320 米
建筑层数	66 层
地址	东大名路 501 号
开放时间	10:00—22:00
公共交通	地铁 12 号线国际客运中心站；公交 22、37 路东大名路旅顺路站，公交 19、25 路唐山路新建路站
Architect	Skidmore, Owings and Merrill, ECADI
Built	2017
Building features	Office, Hotel, Retail
Total building height	320m
Total floors	66
Address	501 Dongdaming Rd.
Recommended viewing time	10:00-22:00
Public Transportation	Metro Line 12 International Cruise Terminal Station; Bus 22/37 East Daming Rd./Lushun Rd. Station; Bus 19/25 Tangshan Rd./Xinjian Rd. Station

白玉兰广场与北外滩
Sinar Mas Plaza and the North Bund

北外滩是上海航运业的发祥地。1845年，英商率先于虹口徐家滩（现东大名路、高阳路）一带建造简陋的驳船码头（虹口码头，1864年改建为轮船码头），此后外商、轮船招商局轮番抢滩兴建码头、货栈、船坞等，几经变更合并。至1949年，虹口沿江一带有汇山码头（1872，1913年至1917年重修）、华顺码头（1860，旗昌码头，又名老宁波码头，1883年更名）、公平路码头（1949，原招商局北栈，1980年代为上海海关关栈暨保税仓库，2008年停运）、高阳路码头（合并1845年虹口码头和1875顺泰码头，现为标志性建筑"一滴水"）。

随着码头的兴起，附近通往下海庙（始建于清代乾隆年间，早期渔民奉祀海神的民间神庙，抗战初期毁于炮火，1941年重建）提篮桥一带（现东大名路、海门路、霍山路交汇区域），逐渐由传统集市成为近代商业繁荣地段。周围兴建城市公共设施：工部局监狱（1903，又称提篮桥监狱，后更名上海市监狱、上海市提篮桥监狱，1999年部分改造为上海监狱陈列馆，现上海监狱博物馆），舟山公园（1917，1948年更名霍山公园，1966年关闭，1984年重修），工部局巡捕医院（1932，1949年上海市公安警察医院，1959年提篮桥区中心医院，1960年虹口区中心医院，1994年上海市中西医结合医院），雷士德工学院（1934，1944年停办，1946年迁入吴淞商船专科学校，后更名上海航务学校，1953年招商局医院、上海海院职工疗养院，1955年上海海员医院，2022年完成整体修缮）。

开埠早期，俄国人多于虹口外滩一带居住。二战期间，大批德、奥犹太难民经海路抵达上海，一部分人聚居在早先俄国犹太人改建的摩西会堂（1927，2007年建上海犹太难民纪念馆）周围的弄堂里避难，开启"第二故乡"新生活，开设理发、鞋帽、服装、五金、面包、小吃等商店，以及酒吧、夜总会、露天屋顶花园等娱乐场所，坊间称为"小维也纳"。霍山公园和摩西会堂成为他们的休闲场所和精神家园。战争结束后至60年代，犹太人陆续离境，留下"维也纳皮鞋店"等，后来成为上海的"中华老字号"。90年代开始，陆续有当年犹太难民返沪寻根，街区内的建筑以清水红砖和连续拱券式样的安妮女王复兴风格为特色。

1990年以前，提篮桥地区的商业繁荣程度与徐家汇不分高低，丰富的商业业态、接壤"大杨浦"的地理优势以及与外滩之间便捷的公交线路，使其成为市民购物"兜马路"的选择之一。1984年提篮桥商业闹市中心建成人行立交桥（2004年拆除），1988年建成27层的远洋宾馆（初名亚洲宾馆，顶楼旋宫为上海首家旋转餐厅）高度超越此前于1978年建造的19层楼大名饭店。90年代之后，该地区日渐萧条。

2002年上海启动黄浦江两岸综合开发建设。2008年在上海船厂（1865年英商耶松船厂，1936年英联船厂，上海近代第一家船舶修造厂）

原址上建成上海港国际客运中心，为北外滩地区注入新的活力。此后在滨江和提篮桥地区，建起以白玉兰广场（2017）、远洋大厦（2000）、外滩W酒店（2017）、北外滩来福士双子塔（2021）等标志性高层建筑，形成与百年外滩、陆家嘴相呼应的天际轮廓线。2020年，一条以反映上海百年码头文化历史的"长廊"——码头文化露天博物馆在江边建成。

毗邻北外滩的杨浦滨江，曾是上海近代最大的工业基地，有着丰厚的工业遗产。其中1910年兴建的杨树浦发电厂（1929，又称江边电站，上海电力公司，后为杨浦发电厂，2011年终止发电）一度垄断整个上海市的电力，于1941年建105米高的烟囱，是当时上海最高的构筑物，现存于上海历史博物馆。1949年以后，杨树浦一带工厂和公用事业陆续由政府接管，经历从五六十年代的公私合营、公有制改造到80年代之后的国有企业改革，传统产业结构面临转型与升级。

乘着世博会的东风，2010年打造的上海国际时尚中心（1921年日商裕丰纱厂，1945年上海第十七纺织厂，1950年国营上海第十七棉纺织厂，2007年迁至江苏大丰）是杨浦滨江从工业岸线向滨江生活岸线转型的华丽开篇。2013年底，上海决定开发杨浦滨江生态岸线，2019年杨浦滨江南段建成开放。同年11月，习近平总书记在杨浦滨江公共空间考察期间，提出"人民城市人民建，人民城市为人民"的重要理念，赋予上海建设新时代人民城市的新使命。从工业锈带到生活秀带，再到发展绣带，北外滩和杨浦的滨江岸线正循着城市发展脉络，成为有历史厚度、城市温度以及社区活力的滨水公共空间。

The North Bund is the birthplace of shipping industry in Shanghai. Commercial firms from England started to build simple and rough barge piers at Xujiatan in Hongkou (now Dongdaming Road and Gaoyang Road). Henceforth, lots of foreign business bodies and the China Merchants snatched shoals and took turns to build piers, docks, and warehouses. After several merges, a few of major piers were stabilized.

The Xiahaimiao Temple (temple for fishermen to revere and worship the sea god) was built in Qianlong age of the Qing Dynasty. It was destroyed during the War of Resistance against Japanese Aggression and reconstructed in 1941. With the thrives of piers and docks, the area nearby the temple (the intersection of East Daming Road, Haimen Road, and Huoshan Road in present time) had grown to a booming business zone from a traditional country fair. Meanwhile, some public service facilities had been put up like the SMC Prison (now Tilanqiao Prison), Zhoushan Park (now Huoshan Park), Shanghai Municipal Police Hospital (now Shanghai

TCM-Integrated Hospital), and the former building of the Lester School and Henry Lester Institute of Technical Education.

Soon after the opening of Shanghai Port, Russians settled down on the Bund near Hongkou. During World War II, so many Jewish refugees from Germany and Austria fled to Shanghai by sea. Some of them gathered in the Ohel Moishe Synagogue refurbished by the Jews from Russia. They opened varieties of business from barbershops, clothing departments, hardware stores, bars, clubs, rooftop leisure gardens to bakeries. As they started the new life in "the second homeland," Huoshan Park and Ohel Moishe Synagogue had become the social spots where they could relax and seek peace in minds. The area where they dwelled was once called "Little Vienna." After the war till the 1960s, Jews had left Shanghai in succession; the shops left behind had become famous "China Time-Honored Brand" in Shanghai. At the beginning of the 1990s, some Jews came back here for reminiscences. Styled with the Queen Anne Revival Architecture, the buildings featured red bricks and multiple-layered arches in this area.

Before 1990, the accomplishments and prosperousness in the area of Tilanqiao and Xujiahui were evenly matched. With broad selections of merchandises, various markets, geographical advantage in connection with "the Great Yangpu," and convenient transportation to the Bund, Tilanqiao was one of the hot spots for shopping and wandering. A pedestrian overpass was constructed in Tilanqiao downtown in 1984. 27-story Ocean Hotel featured the first revolving restaurant in Shanghai was built in 1988. Its height exceeded 19-story Daming Hotel completed in 1978. Unfortunately, Tilanqiao business zone started to fall in depression after the 1990s.

In 2002, the comprehensive plan of exploitations and innovations on the shores of Huangpu River was launched. The completion of Shanghai Port International Cruise Terminal, on the ole site of Shanghai Shipyard in 2008 generated the productive forces in the Bund's developments. The iconic tall buildings like Sinar Mas Plaza (2017), Sino-Ocean Tower (2000), W Shanghai-The Bund (2017), Twin Tower of Raffles City The Bund (2021) along the riverside and Tilanqiao sketch out the skyline together with the architectures in the Bund and Lujiazui. In 2020, Shanghai Docks Heritage Museum was finished on the shore of Huangpu River. It is an art gallery displays and illustrates the "culture of the Shanghai Port" over one hundred years.

The bank of Huangpu River in Yangpu District is right next to the North Bund. It was the largest industrial ground in modern Shanghai. It is rich in legacy of industries. Yangshupu Power Plant of SMC (Riverside Power Station) built in 1910 was one of the examples. The plant had controlled power system and electricity supply in the entire city for a time. The 105-meter-tall chimney on the plant was the city's tallest structure in 1941. It was closed in 2011. After 1949, the factories and public services in Yangshupu were taken over by the government. Through the public-private partnership process and public ownership revamping between the 1950s and 1960s and reform of state-owned enterprises after the 1980s, the traditional structure of manufacture was facing necessary transformations and upgrades.

Driven by the World Expo, Shanghai Fashion Center was put up on the Yangpu Riverside in 2010. The site used to be Shanghai the 17th Cotton Textile Factory, its predecessor was a yarn factory opened by Japanese company in 1921. The factory was moved to Dafeng in Jiangsu Province in 2007. The rising of Shanghai Fashion Center indicated that the characteristics of Yangpu had begun the transition from industry to liveliness. At the end of 2013, Shanghai Municipal decided to start ecosystem project at the waterfront in Yangpu. The southern segment in ecosystem was completed and open to the public in 2019. In November of the same year, President Xi Jinping proposed an important philosophical concept while he was observing and inspecting the progression in Yangpu. This concept, "The city is built by the people and for the people," became the great mission in the new era of the development of Shanghai.

Transiting from the industrial rust (锈Xiu) to life show (秀Xiu), and thriving to refinement and elegance (绣Xiu), the North Bund and shores along the Huangpu River in Yangpu have been growing into an energetic waterfront public space with rich history and charming urban fashion.

白玉兰广场与陆家嘴
Sinar Mas Plaza and the Lujiazui

白玉兰广场 | Sinar Mas Plaza

人人屋

人人屋党群服务站是杨浦滨江南段公共空间的一处滨水驿站,项目所在地为祥泰木行(1902)的旧址。建筑为市民提供休憩驻留、医疗救助、全息沙盘、微型图书室等服务,体现以人为本、为人民服务的理念。驿站所在的滨水区段为塔吊演艺区,当该区域举行文艺表演时,也能够作为舞台的辅助用房。建筑采用钢木结构,并以"人"字形杆作为基本单元进行搭建,遂取名为人人屋。

The Yangpu Riverside "Ren Ren Wu"

"Ren Ren Wu" is the building name of Yangshupu Station in the south section of Yangpu Riverside. The project is located at the former site of the Xiangtai Wood Firm (started in 1902). The building provides residents with lodging, relaxing, and other conventional services, as well as medical assistance, a holographic sand table, and a miniature library. All the service facilities are based on the concept of "people-oriented and serving the people." The waterfront section of the post station is the tower crane performance area. It can also be used as a main stage's auxiliary room when performing shows held in this area. The building adopts a steel and wood structure and is built with "herringbone" rods as the basic unit. This is how the name of "Ren Ren Wu" (means Pavilion) was received.

虹桥交通枢纽：虹桥商务区的崛起
Hongqiao Transportation Hub: The rise of Hongqiao CBD

集机场、高速铁路、市域铁路、地铁站、长途汽车站、公交站、出租车候车等于一体的综合交通枢纽。以架次繁多的飞机作为"移动的天街"，回眸与展望这座城市——海纳百川，天容万物。

Integrated with airport, high-speed railways, urban railroads, subway stations, coach stations, bus stations, and taxi service, it is the most detailed and sophisticated. We may take rides on the airplanes, which are "moving streets in the sky," to overlook Shanghai. This city is a tremendous melting pot, inclusive and diverse.

建筑师	华东建筑设计研究院有限公司
建成时间	2009 年
建筑功能	机场、高铁站、长途汽车站、轨道交通、零售
建筑高度	41.2 米
建筑层数	3 层
地址	申贵路 1500 号（上海虹桥站），申达一路 1 号（上海虹桥国际机场 2 号航站楼）
开放时间	全天
公共交通	地铁 2、10、17 号线虹桥火车站，公交虹桥枢纽 7、10 路虹桥西交通中心站，公交 173 路、闵行 28 路申虹路舟虹路站
Architect	ECADI
Built	2009
Building features	Airport, Railway Station, Long-distance Bus Station, Subway Station, Retail
Total building height	41.2m
Total floors	3
Address	1500 Shengui Rd. Shanghai Hongqiao Station, 1 Shenda Rd. (1) (Terminal 2 Shanghai Hongqiao International Airport)
Recommended viewing time	All day
Public Transportation	Metro Lines 2/10/17 Hongqiao Railway Station; Bus Hongqiao Hub No.7/10 Hongqiao West Transportation Center Station; Bus 173/Minhang No.28 Shenhong Rd./Zhouhong Rd. Station

虹桥机场远眺陆家嘴
From Hongqiao Airport to Lujiazui

1901年工部局越界筑路，东起海格路（现华山路）杨家库（现广元路一带）南洋公学（1896，1921年更名交通大学）校门南侧，西拐新泾至横沥港，为煤屑路和泥路（初建时今程家桥以西路段称佘山路），虹桥乡位于路南。1920年代初《南洋大学学生生活》记载，虹桥路在学校南边，路旁有小河及一座叫"虹桥"的小桥。彼时虹桥路东段与徐家汇相距不远，成为较早城市化的路段。1921年，北洋政府在虹桥路以西上海县和青浦县交界处建成军用机场。

20世纪三、四十年代，虹桥路沿线修建了一批乡间别墅、休闲场所、医疗机构、学校、仓库、苗圃等。1949年以后，部分改为城市公共设施：西郊公园（1954，1980年更名上海动物园；原址为1916年英商高尔夫球场），龙柏饭店（1986，原址保留1932年沙逊别墅），上海舞蹈学校（1961年迁此，2016年建成国际舞蹈中心），上海市盲童学校（创办于1912年，1928迁至此，1952年政府接管），宋庆龄陵园（1984，前身为1916年万国公墓）。此外，由华人建筑师设计的虹桥疗养院（1934，曾用作中福会托儿所、上海船舶制造学校、虹桥医院等，现上海血液中心）被建筑史学家认为是上海近代建筑史上具有代表性的现代主义建筑之一。

1950—1951年，机场于原址重建，60年代扩建国际航站楼，并取代龙华飞机场成为上海唯一的航空港，1971年由军民合用改为民用机场。此后分别于1984年、1991年、1996年再行改扩建工程。1986年，古北路至虹桥机场之间的虹桥路段拓宽，在90年代延安路高架建成以前，是从机场进市区的"国宾道"。90年代末虹桥路东段更名广元西路，改道直抵徐家汇。2018年，虹桥路向西延伸至机场内部。

临近机场的虹桥经济技术开发区从1979年开始规划，是改革开放初由国务院批准的国内首批14个经济技术开发区之一，以外贸为引领，集展览贸易、涉外办公、商业居住等功能于一体的新型商贸区，曾被评选为"90年代上海十大新景观"之一。1988年和1990年建成的虹桥宾馆和银河宾馆，是开发区最早投入的项目，为同一基地内的高层建筑姊妹楼（现为锦江郁锦香和新业中心）。1996年，上海人民广播电台从北京东路外滩迁至开发区。

2009年虹桥综合交通枢纽建成，虹桥商务区也应运成立，就此拉开"大虹桥"建设的序幕。早在1959年，虹桥地区就建有上海农业展览馆（1997年新馆），1983年成立上海历史文物陈列馆（1991年更名上海市历史博物馆，1999年关闭，2018年于南京西路原跑马总会大楼重新开放），1996年上海少儿博物馆开馆，1999年建成上海世贸商城，含常年展贸中心和世贸展馆。2014年9月建成国家会展中心（上海），同年11月，凌空SOHO（2019年更名Sky Bridge HQ 天会）竣工，16条空中走廊以前卫的造型令人惊艳。

在建的市域铁路机场联络线即将连接虹桥、浦东两大国际机场，高铁和航空技术的提升进一步奠定了虹桥交通枢纽作为国内外双循环重要节点的地位，未来的虹桥商务区将承载城市的梦想与希望，建设成为面向全球和引领长三角区域高质量一体化发展的国际开放枢纽中央商务区。

In 1901 the SMC constructed a road crossing its territory border. The east end of the road started from Avenue Haig (now Huashan Road), Yangjiashe (now Guangyuan Road), and the Nanyang Mission College (founded in 1896, renamed as Jiaotong University in 1921), and its west end bent into Xinjing Town. The road was covered by cinders and muds. The segment to the west of today's Chengjiaqiao was called Sheshan Road. Hongqiao Town located on the south side of the road. As stated in *Nanyang Mission College Students*, Hongqiao Road was to the south of the school. There was a small creek next to Hongqiao Road. On the water stood a small bridge named "Hongqiao." The reddish bridge and lush trees along the road complemented each other just as a gorgeous picture of nature's beauty. Not far from Xujiahui at that time, the east segment of Hongqiao Road represented the urbanization in the early age. In 1921, the Beiyang Government built a military airport to the west of Hongqiao Road which was at the junction of Shanghai County and Qingpu County.

Between the 1930s and 1940s, country houses, clubs, medical institutions, schools, warehouses, and nurseries were constructed along the Hongqiao Road. The area with bucolic scenery were fascinated by these modern-fashioned facilities. Some of these facilities were transformed into the public services after 1949. Xijiao Park was established in 1954 on the site of English Golf course from 1916, and later was renamed as Shanghai Zoo in 1980, Cypress Hotel was built in 1986 on the property of former Villa Sassoon which was remained in its original conditions. Others like Shanghai Dance School and the Shanghai Blind Children School were move to this area in different years. Shanghai Hongqiao Sanatorium, designed by Chinese architects, was recognized by architectural historians as one of the buildings embodied modernism in Shanghai architecture history.

The airport was reconstructed between 1950 and 1951. An international terminal building was appended in 1960. By supplanting the Longhua Airport, the Hongqiao Airport became the only air harbor in Shanghai. In 1971, the airport ended its the military duties and functioned just as a civil airport. There were 3 expansions in 1984, 1991, and 1996. The section of Hongqiao Road between Gubei Road and Hongqiao Airport was broadened in 1986. Before Yan'an Elevated Road was built, Hongqiao Road was "the

Passage for State Guests" from the airport to the urban area. At the end of the 1990s, the eastern section of Hongqiao Road was renamed West Guangyuan Road and rerouted to Xujiahui directly. In 2018, Hongqiao Road was extended all the way to the airport interiors.

Hongqiao Economic Development Zone near the Hongqiao Airport was designed and planned in 1979. It was one of the 14 Economic Development Zones approved by the State Council at the early period of Reform and Opening-up. Focusing on foreign trades, Hongqiao Economic Development Zone is an inclusive complexity of international exhibition, foreign trade & business, and commercial residences. The Rainbow Hotel built in 1988 and Galaxy Hotel built in 1990 were the projects executed foremost. Two hotels are the sister buildings on the same field. In 1996, Radio Shanghai was moved into Hongqiao Economic Development Zone from East Beijing Road.

Hongqiao CBD was established as the Hongqiao transportation hub formed in 2009. This was the prelude of the constructions in "Great Hongqiao." Shanghai Agricultural Exhibition Hall and Shanghai Historical and Cultural Relics Exhibition Hall had already existed in 1959. The new exhibition hall of Shanghai Agricultural Exhibition Center was completed in 1997. Shanghai Historical and Cultural Relics Exhibition Hall was renamed as Shanghai History Museum in 1997. After the relocation on the site of Racing Club on Nanjing Road, Shanghai History Museum reopened to the public. Other major constructions, like Shanghai Children's Museum (1996) and Shanghai Mart (1999). In 2014, following the NECC (Shanghai) and the Sky Bridge SOHO had successively been completed.

Nowadays, Airport Link Line under the construction will connect both Hongqiao and Pudong International Airports. High-speed railways and advanced aviation technologies have positioned Hongqiao Transportation Hub on the essential point of domestic and international double loops. In the future, carrying the dreams and hopes of the Shanghai City, Hongqiao CBD will be developed as a commercial hub, not only leading the high-quality integrated developments of the Yangtze River Delta region, but also open to the globe.

大虹桥
Greater Hongqiao Area

国家会展中心(上海)

国家会展中心(上海)于 2014 年 9 月竣工,总建筑面积超 150 万平方米,是中国国际进口博览会的主场馆,集展览、会议、商业、办公、酒店等多种业态为一体。四叶草造型的主体建筑入选 2019 年上海新十大地标建筑,是目前世界上最大规模的单体会展建筑。

National Exhibition and Convention Center (Shanghai)

Completed in Sept. 2014, with a total floor area of more than 1.5 million square meters, the NECC (Shanghai) is the main venue of China International Import Expo. The muti-functional complex with four-leaf clover in plan makes it a unique landmark overlooking from the air. It is the largest convention and exhibition complex in the world.

后记 迈向人民的垂直城市

上海作为中国最具代表性的近现代摩天都市，中心城区分布有不同时期、数量巨大、类型丰富，并颇具代表性的高层建筑。一方面，建筑始终是上海的第一张名片，曾被称为"世界建筑博览会"，各时期的高层建筑无论从科学技术、文化艺术还是功能设置上都可以比肩全球。另一方面，上海城市发展和高层建筑的建造相辅相成，在城市形态的变迁中可见现代垂直城市发展演变的缩影。近年来，越来越多的高层建筑增设空中公共空间，提供了俯瞰或远眺城市的机会与各种视角，人们得以切身体验上海这座城市的多元性、历史性和文化性。

我们记录梳理上海中心城区代表性空中公共空间及其高空视野，展现城市发展过程中的有趣片段；对城市上空行走带来的体验以及所见景象的历史成因作出解读；借助"空中行走"提供"阅读城市"的立体视角，为读者带来对熟悉城市的多维时空认知，以期把"摩天都市"的物质标签与"人民城市"的日常生活更紧密地联系起来，让公众体验到现代垂直城市中公共生活的更多可能性。

尽管上海的垂直城市空中公共空间已颇具规模，但市民知之甚少，仍处在自发生长阶段。一是空中公共空间分布不均，许多重要区域仍处在"盲区"；二是高空视野缺乏统筹规划，缺少层次、利用不充分和重复度较高；三是开放空间的公共性普遍较弱，以高消费场所为主。

打造"摩天都市"，创建"人民城市"，是上海一直以来的重要目标，亦是今后城市立体化发展的关键。空中公共空间体系或将成为上海垂直城市建设的重要一环，更多空中公共空间的涌现以及整体性规划的导控，势必有助于解决高密度城市中心区公共空间不足的问题，并将整个上海的城市空间连接成更为紧密的整体，从而实现人民的垂直城市。

在五年的成书过程中，我们被一代又一代上海城市建设者的努力深深触动，他们在城市规划建设中，在城市遗产保护中，在城市更新过程中，兼收并蓄、博采众长、缜思畅想，激发了我们作为后辈的责任、勇气和担当。

需要说明的是，"新沪上十八景"是从各个区域众多案例中筛选出来的，兼具空中视野与场所体验，并向公众开放的空中观景最佳去处。但由于能力与时间的制约，难免存在疏漏，希望与读者们一起继续完善。期待在未来与更多空中公共空间的惊喜邂逅。

王桢栋，冯宏，杨鹏宇，白雪君
于同济园

POSTSCRIPT Towards a Vertical City for the People

Shanghai, as the most representative modern skyscraper city in China, its central urban area is characterized by different periods, a great number and diversity of quite representative high-rise buildings. On one side, architectures have always been Shanghai's first business card. The city is widely known as the "World Architecture Expo." Its high-rise buildings in various periods can be compared with the rest of the world in terms of science and technology, culture and art, and functional settings. On the other side, Shanghai's urban development and the construction of high-rise buildings complement each other. Reading the history of its urban form can be seen in the epitome of the evolution of modern vertical urban development. In recent years, more and more high-rise buildings in Shanghai have begun to add aerial public spaces, and the view from these buildings offers a way to experience the diversity, history and culture of Shanghai.

First, we note and sort out the typical public aerial spaces in the current downtown area and the views from those tall buildings to show the interesting fragments during the different periods of the urban development. Secondly, we academically yet leisurely interpret the experiences from walking "over" the city and introduce the historical backgrounds of the sceneries. Finally, we provide a three-dimensional perspective of "reading the city" via "air walking" to help readers get a better understanding on a concept of chrono-spatial in multi dimensions. With our work in three aspects stated above, we have set the conceptual system and provided details of the various aerial public spaces and their views, which involve the history and culture of the city, in downtown Shanghai. We are expecting to see the close links between the label of "skyscraper city" and the true daily life of "people's city." We hope that more people can enjoy the social life in the public and experience the differences between traditional cities and modern vertical cities. We also hope that our work will interest more people not only to enjoy the urban aerial public spaces but also start to explore this subject spontaneously and actively.

At the same time, we have noticed that, even though the constructions of vertical urban air public space in Shanghai has reached a considerable scale, a large amount of the residents in the city are still lack of acknowledgement. The growth of such constructions and popularity are quite spontaneous rather than orderly and systematical. Roughly, we are now facing the following three problems: First, the distribution of aerial public space is uneven, and many important areas are still in the "blind zone" in an aerial vision. Second, the lack of planning and frame work caused poor gradation of scenic views, boring repetition, and insufficient utilization. Third, the publicity of open space is generally weak, because its functions are mainly high-value consumption businesses such as upscale bars and restaurants.

Building a "skyscraper city" and creating a "people's city" have always been important goals in Shanghai, and these are also the keys to the three-dimensional development of the city in the future. As the urban constructions moving forward and metropolitan management continuous improving, the aerial public space system may become an important part of Shanghai's vertical city construction. With the emergence of more aerial public space and the more guided and controlled overall planning, it may be possible to effectively solve the problem of insufficient public space in the central area of high-density cities. We will see how the aerial public space system connects the urban spaces throughout the city Shanghai into an entity in which the concept of vertical city becomes a true and vivid sense to people.

In the period of five years working on this book, we have been deeply touched by the efforts of generations of Shanghai city builders, including many alumni of Tongji University. They have collected valuable information, given deep thoughts, been engaged in discussions, and learned so much from urban planning and constructions, urban heritage protection programs, and the process of urban regeneration. The hard work of the predecessors has inspired the young generations of responsibilities and encouraged them to undertake the tasks in the future.

It should be noted that the "New 18 Scenic Spots in Shanghai", selected by the authors from many cases in various regions, combines the characteristics of unique aerial views and on-spot experiences and currently open to the public. However, due to the constraints of ability and time, it is inevitable that there will be omissions, so we hope to continue to improve our catalogue of such spaces through the interaction with readers. More openings and presentations of urban air spaces in cities are expected with no doubts.

Wang Zhendong, Feng Hong, Yang Pengyu, Bai Xuejun
on Tongji University Campus

黄浦江越江基础设施基本信息
Basic Information of Huangpu River Crossing Infrastructure

设施 Infrastructure	地址 Address	通车时间 Open to Traffic
牡丹江路/富锦路，S20外环线 Mudanjiang Rd./Fujin Rd., S20 Outer Ring Ewy.		
泰和路/同济路，三岔港 Taihe Rd./Tongji Rd., Sancha Port, component of S20		2003.06.21
长江西路/军工路，港城路/双江路 Changjiang Rd.(W)/Jungong Rd., Gangcheng Rd./Shuangjiang Rd.		2016.09.10
国帆路站，双江路站 Guofan Road Station, Shuangjiang Road Station		2019.12.28
翔殷路/军工路，浦东北路/五洲大道 Xiangyin Rd./Jungong Rd., Pudong Rd.(N)/Wuzhou Ave.		2005.12.31
周家嘴路/内江路，东诸路/张杨北路 Zhoujiazui Rd./Neijiang Rd., Dongjing Rd./Zhangyang Rd.(N)		2019.10.31
复兴岛站，东陆路站 Fuxing Island Station, Donglu Road Station		2020.12.26
军工路，金桥路/张江路 Jungong Rd., Jinqiao Rd./Zhangjiang Rd.		2011.01.28
宁国路/黄兴路，罗山路立交/杨高中路 Ningguo Rd./Huangxing Rd., Luoshan Rd. IC/Yanggao Rd.(M)		1993.10.23
江浦路/龙江路，民生路/商城路 Jiangpu Rd./Longjiang Rd., Minsheng Rd./Shangcheng Rd.		2021.09.30
丹阳路站，昌邑路站 Danyang Road Station, Changyi Road Station		2021.12.30
大连西路/霍山路，东方路 Dalian Rd.(W)/Huoshan Rd., Dongfang Rd.		2003.09.29
提篮桥站，浦东南路站；三林站，华泾西站 Tianqiao Station, South Pudong Road Station; South Sanlin Station, West Huajing Station		在建中 Under construction
南浦大桥站，杨树浦路站，浦东大道站 Nanpu Bridge Station, Tangqiao Station, Yangshupu Road Station, Pudong Avenue Station		2007.12.29
海伦路/海拉尔路，银城东路/银城中路 Hailun Rd./Hailaer Rd., Yincheng Rd.(E)/Yincheng Rd.(M)		2010.03.26
外滩中山东一路300号，滨江大道2789号 No.300 Zhongshan Rd.(E-1), No.2789 Binjiang Blvd.		2000.10

隧道 Tunnel ／ 桥梁 Bridge ／ 地铁 Metro

- 郊环隧道 Jiaohuan Tunnel
- 上海外环隧道 Shanghai Outer Ring Tunnel
- 长江路隧道 Changjiang Road Tunnel
- 地铁10号线 Metro Line 10
- 翔殷路隧道 Xiangyin Road Tunnel
- 周家嘴路隧道 Zhoujiazui Road Tunnel
- 地铁12号线 Metro Line 12
- 军工路隧道 Jungong Road Tunnel
- 杨浦大桥 Yangpu Bridge
- 江浦路隧道 Jiangpu Road Tunnel
- 地铁18号线 Metro Line 18
- 大连路隧道 Dalian Road Tunnel
- 地铁19号线 Metro Line 19
- 地铁4号线 Metro Line 4
- 新建路隧道 Xinjian Road Tunnel
- 外滩观光隧道 Bund Sightseeing Tunnel

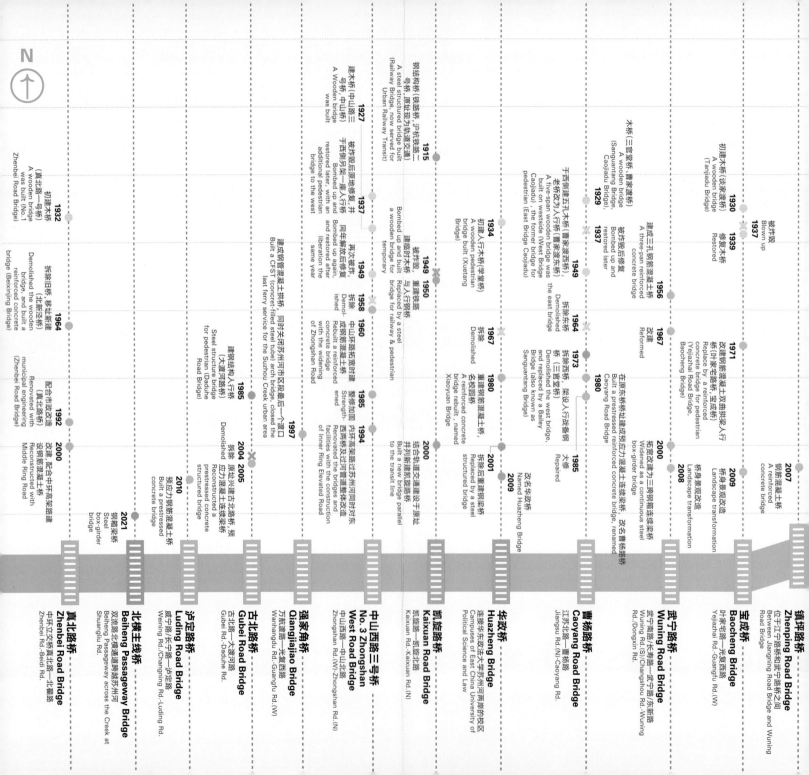

延安路 Yan'an Line & 南线 South Line

Date	Location (Chinese)	Location (English)	Tunnel/Line
1971.06.10	中山一路/打浦路, 耀华路/浦东大道	Zhongshan Rd.(S-1)/Dapu Rd., Yaohua Rd./Pudong Ave.	打浦路隧道 Dapu Road Tunnel
1989.12.??	延安东路/福建路, 世纪大道/银城中路	Yan'an Rd.(E)/Fujian Rd.(M), Century Ave./Yincheng Rd.(M)	延安东路隧道 Yan'an Road E. Tunnel
1991.12.01	中山南路/中山南一路, 龙华西路, 雪野路/浦东南路	Zhongshan Rd.(S)/Zhongshan Rd.(S-1)/Longhua Rd.(W), Xueye Rd./Pudong Rd.(S)	南浦大桥 Nanpu Bridge
1997.06.24	龙吴路立交, 济阳路跨线过江通道	Longwu Rd. IC, Jiyang Rd. IC, component of S20	徐浦大桥 Xupu Bridge
2003.06.28	鲁班路立交, 耀华路/济阳路立交	Luban Rd. IC, Yaohua Rd./Jiyang Rd. IC	卢浦大桥 Lupu Bridge
2004.09.29	西藏南路站, 中华艺术宫站	South Xizang Road Station, China Art Museum Station	西藏南路隧道 Xizang Road S. Tunnel
2007.12.29	西藏南路/中山南路, 龙华东路, 雪野路/浦东南路	Xizang Rd.(S)/Zhongshan Rd.(S)/Longhua Rd.(W), Xueye Rd./Pudong Rd.(S)	地铁8号线 Metro Line 8
2009.11.20	人民路/淮海东路, 东昌路	Renmin Rd./Huaihai Rd.(E), Dongchang Rd.	人民路隧道 Renmin Road Tunnel
2009.12.05	龙华中路站, 后滩站	Middle Longhua Road Station, Houtan Station	地铁7号线 Metro Line 7
2009.12.31	复兴东路/光启路, 张杨路/崂山西路	Fuxing Rd.(E)/Guangqi Rd., Zhangyang Rd./Laoshan Rd.(W)	复兴东路隧道 Fuxing Road E. Tunnel
2010.02.11	中山一路/打浦路, 耀华路/浦东大道 复线	Zhongshan Rd.(S-1)/Dapu Rd., Yaohua Rd./Pudong Ave. Double Tunnel	打浦路隧道 Dapu Road Tunnel
2010.04	马当路站, 世博会博物馆站, 世博大道站	Madang Road Station, World Expo Museum Station, Shibo Avenue Station	世博园专用隧道 Expo Park Special Tunnel
2010.04.15	云锦路/龙耀路, 成山路/长青路	Yunjin Rd./Longyao Rd., Chengshan Rd./Changqing Rd.	龙耀路隧道 Longyao Road Tunnel
2010.04.20 → 2020.11.01	世博专线地铁 (后作为地铁13号线的一段)	Metro Expo Line (as a part of the later Metro Line 13)	地铁13号线/世博专线地铁 Metro Line 13/Metro Expo Line
2011.02.01		Open to Traffic 通车	
2013.08.31	龙耀路站, 东方体育中心站	Longyao Road Station, Oriental Sports Center Station	地铁11号线 Metro Line 11
2015.12.10			地铁13号线 Metro Line 13
2021.12.30			地铁14号线 Metro Line 14
在建中 Under construction	龙水南路/中环立交, 济阳路立交	Longshui Rd./Middle Ring IC, Jiyang Rd. IC	龙水南路隧道 Longshui Road S. Tunnel
在建中 Under construction	虹梅路/中环立交, 济阳路立交	Hongmei Rd./Middle Ring IC, Jiyang Rd. IC	上中路隧道 Shangzhong Road Tunnel
在建中 Under construction	联系虹桥国际机场和浦东国际机场	Urban rail transit between Shanghai Hongqiao International Airport and Shanghai Pudong International Airport	机场联络线 Airport Link
在建中 Under construction	景东路/银都路, 浦锦路/芦恒路	Jingdong Rd./Yindu Rd., Pujin Rd./Luheng Rd.	银都路隧道 Yindu Road Tunnel

豫园站, 陆家嘴站 — Yuyuan Garden Station, Lujiazui Station

小南门站, 商城路站 — Xiaonanmen Station, Shangcheng Road Station

Suzhou Creek Bridges Timeline

福建路桥 Fujian Road Bridge
福建中路/北京东路、福建北路/天潼路
Fujian Rd.(M)/Beijing Rd.(E)/Fujian Rd.(N) /Tiantong Rd.

- **1864** 建木桥（原潘龙桥）A wooden bridge on a stone gate (Pentanlong Bridge) was built
- **1875** 重建七孔木桥（名老闸桥）A seven-pan wooden bridge (Laozha Bridge) was rebuilt
- **1946** 再建木桥，更名福建路桥 A wooden bridge was built, named Zhejiang Road Bridge
- **1968** 建钢筋混凝土桥，最早的双曲拱桥 A reinforced concrete bridge was built. It is the first hyperbolic arch bridge in China
- **2001** 拆除 Demolished
- **2004** 重建 Rebuilt

浙江路桥 Zhejiang Road Bridge
浙江中路/厦门路、浙江北路/曲阜路
Zhejiang Rd.(M)/Xiamen Rd./Nansuzhou Rd.-Xizang Rd.(N)/Qufu Rd.

- **1880** 石桥 Stone bridge
- **1887** 重建木桥（老闸桥）Replaced by a wooden bridge
- **1908** 改建鱼腹式钢桁架桥（老垃圾桥）A steel Fishbelly bridge (Old Laji Bridge)
- **1942** 更名浙江路桥 Named Zhejiang Road Bridge
- **1975** 大修加固 Overhauled and restored
- **2015** 维修复位 Repaired and replaced

西藏路桥 Xizang Road Bridge
西藏中路/厦门路、西藏北路/曲阜路
Xinqiao Rd.-Xizang Rd.(M)/Xiamen Rd.-Xizang Rd.(N)

- **1853** 初建木桥（泥城桥）The wooden Bridge (Nicheng Bridge) was built
- **1899** 改建木桥，新位置建 Rebuilt
- **1924** 改建钢筋混凝土桥（文名垃圾桥）新 A reinforced concrete bridge, New Laji Bridge was built
- **1942** 加固加宽，更名西藏路桥 Reinforced and widen (called the Xizang Road Bridge)
- **1948** 重建木桥 Rebuilt a wooden bridge
- **1985** 拆除木桥改建钢筋混凝土桥 Demolished, and replaced by a reinforced concrete bridge
- **1997** 拆除 Demolished
- **1999** 重建 Rebuilt

乌镇路桥 Wuzhen Road Bridge
新闸路-大统路
Xinzha Rd./Wuzhen Rd.-Datong Rd.

- **1735** 石闸上浮桥，为吴淞江（苏州河）上第一桥 Originally a wooden bridge on a stone gate
- **1862** 改建吊桥 A suspension bridge was rebuilt
- **1929** 建六孔木桥 A six-pan wooden bridge
- **1937** 淞沪会战期间被毁 Destroyed during the War of Resistance against Japanese Aggression
- **1948** 重建木桥，更名乌镇路桥 Rebuilt a wooden bridge
- **1968** 开造加宽 Overhauled and restored
- **1999** 建成钢管混凝土拱桥 (concrete-filled steel tube) arch bridge
- **2002** 拆除 Demolished replaced

新闸桥 Xinzha Bridge
新闸路-大统路
Xinqiao Rd.-Datong Rd.

- **1897** 建木桥，名新闸桥 A wooden bridge (Xinzha Bridge)
- **1916** 建成木桩基钢桁梁桥 A steel truss wooden pile bridge
- **1927** 大修后禁止重车通行 Heavy trucks was forbidden
- **1999** 拆除旧桥，改建成钢桁梁跨越结构人行道上桥架（预制土桁通机动车结构） Demolished and replaced by a steel structure pedestrian bridge (with the design capacity of motor vehicle)
- **2010** 综合改造完成机动车通行 Remodeled and upgraded for both pedestrian & vehicle

成都路桥（南北高架桥）Chengdu Road Bridge
南北高架路南越苏州河段
N-S Elevated Rd. across the Suzhou Creek

- **1994** 配合南北高架路建设，在原址东侧建设新桥，迁建原应力混凝土桥 Built as an attached construction of N-S Elevated Rd. (the widest bridge on the Suzhou Creek), on the east side of the original site, the new Hengfeng Road Bridge with the construction of Shanghai Railway Station

恒丰路桥 Hengfeng Road Bridge
石门二路-恒丰路
Shimener Rd.-Hengfeng Rd.

- **1903** 初建木桥（汇旗桥）大桥（新纳贩厂桥）新 Wooden bridge (Huitong Bridge, Shanbanchang Bridge, Markham Road Bridge)
- **1914-1917** 因安全隐患拆除修建 Demolished then rebuilt
- **1927** 重建 Rebuilt
- **1948** 建筑五孔钢筋混凝土系梁桥（又称淮安路桥）Built a five-span reinforced concrete bridge (Huai'an Road Bridge)
- **1987** 配合路新达恒越苏州河侧建设原址东侧建新桥现在北侧应力混凝土桥 Built on the east side of the old bridge, renamed original site, the new Hengfeng Road Bridge
- **1989** 老桥拆除 Demolished the old
- **2022** 西侧拱桥增加观光电梯 Observation lift was installed

昌平路桥 Changping Road Bridge
昌平路-恒通路
Changping Rd.-Hengtong Rd.

- **2020** 钢拱桥（苏河之眼）Steel arch bridge (known as "Eye of the Suzhou Creek")

普济路桥 Puji Road Bridge
海防路/淮安路-普济路
Haifang Rd./Huai'an Rd.-Puji Rd.

- **1997** 1997年钢筋混凝土桥，非机动车人行桥 Reinforced concrete bridge, for non-motorized vehicle & pedestrian

长寿路桥 Changshou Road Bridge
长寿路-天目西路
Changshou Rd.-Tianmu Rd.

- **1953** 拆意改造 新建桥，改名长寿路桥 Widened on each side, renamed Changshou Road Bridge
- **1996-1997** 1998 拓宽改建 新建桥，改名长寿路桥
- **2001** 钢筋混凝土桥 A new concrete bridge

昌化路桥 Changhua Road Bridge
昌化路-中潭路
Changhua Rd.-Zhongtan Rd.

- **1929** 建钢筋混凝土桥，是1949年后苏州河上建造的第一座桥，初名长寿桥 A reinforced concrete bridge, which was the first named Changshou Bridge, bridge built on the Suzhou Creek after 1949
- **1951** 建木桥（旧化路桥）A wooden bridge (Guihua Road Bridge)
- **1957** 改名昌化路桥 Renamed Changhua Road Bridge
- **1974** 改建钢筋混凝土桥 Replaced by a reinforced concrete bridge

江宁路桥 Jiangning Road Bridge
江宁路/澳门/光复西路
Jiangning Rd./Aomen Rd.-Guangfu Rd.(W)

- **1903** 初建木桥（造币厂桥）A wooden bridge (Zaobichang Bridge)
- **1949** 重建木排架桥 Rebuilt the wooden bridge
- **1968** 重建预应力钢筋混凝土桥 Replaced by a prestressed reinforced concrete bridge
- **2012** 重建钢筋混凝土桥 Rebuilt a reinforced concrete bridge

苏州河上的桥
Bridges over the Suzhou Creek

● 修建时间 Built Time   ✗ 拆除时间 Demolished Time

● 木桥 Wooden Bridge   ● 石桥 Stone Bridge   ● 钢筋混凝土桥 Concrete Bridge   ● 钢桁架桥 Steel Bridge

1856
木桥 (韦尔斯桥)
A wooden bridge named Willis Bridge

1873
木桥 (公园桥)
Renewed (Garden Bridge, Waibaidu Bridge)

开埠后/After 1843
浮桥
A pontoon bridge

1860
浮桥
A pontoon bridge (named Baida Bridge) was built

1878
建木桥 (重建浮桥,白大桥)
A wooden bridge (named Libaidu Bridge, Baida Bridge) was built

1875
木桥 (三摆渡桥,铁马路桥,天妃宫桥)
A wooden bridge (named Sanbaidu Bridge, Tiemalu Bridge, Tianfeigong Bridge) was built

1873
改为木桥 (头坝渡桥,二白渡桥)
Changed to a wooden bridge (named Toubadu Bridge, Erbaidu Bridge)

1907.12.29
中国第一座钢桥
浮桥 (花园桥)
Rebuilt, the first all-steel bridge in China

1922
改建三孔钢筋混凝土桥 (俗称邮政局桥)
A reinforced concrete bridge (commonly known as the General Post Office Bridge) was rebuilt

1927
改建钢筋混凝土桥
A reinforced concrete bridge had built and named by Zhapu Road Bridge

1927
混凝土桥 (河南路桥)
A concrete bridge (Henan Road Bridge) was built

2008.4-2009.2
整体拆除维修并归位
The bridge was removed for renovation and was restored to its original position after 10 months

2006-2009
重建并拓宽
Rebuilt and widened

1900s ...... 1920s ...... 1940s ...... 1960s ...... 1980s ...... 2000s 2020s

外白渡桥 Waibaidu Bridge
中山东一路/南苏州路 — 大名路/黄浦路/北苏州路
Zhongshan Rd.(E-1)/Nansuzhou Rd./Huangpu Rd./Beisuzhou Rd. - Daming Rd.

乍浦路桥 Zhapu Road Bridge
虎丘路—乍浦路
Huqiu Rd.-Zhapu Rd.

四川路桥 Sichuan Road Bridge
四川中路/南苏州路—四川北路/北苏州路
Sichuan Rd.(M)/Nansuzhou Rd.-Sichuan Rd.(N)/Beisuzhou Rd.

河南路桥 Henan Road Bridge
河南南路/南苏州路—河南北路/北苏州路
Henan Rd.(M)/Nansuzhou Rd.-Henan Rd.(N)/Beisuzhou Rd.

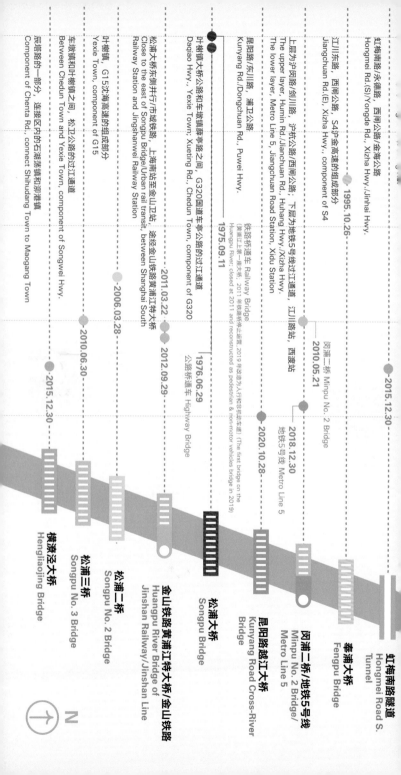

图书在版编目（CIP）数据

上海空中行走地图 = Shanghai Skywalkers：汉英对照 / 王桢栋等著；萤光学社，（美）丹尼尔·萨法里克译. – 上海：上海文化出版社，2022.12
ISBN 978-7-5535-2671-3

Ⅰ.①上… Ⅱ.①王… ②萤… ③丹… Ⅲ.①高层建筑 - 上海 - 图集 Ⅳ.① TU97-64

中国版本图书馆 CIP 数据核字 (2022) 第 241606 号

出版人：姜逸青
责任编辑：江岱
装帧设计：孙大旺 袁之淇

书名：上海空中行走地图
　　　王桢栋 冯宏 尹明 萤光学社 著
　　　萤光学社（美）丹尼尔·萨法里克 译
出版：上海世纪出版集团　上海文化出版社
地址：上海市闵行区号景路 159 弄 A 座 2 楼 201101
发行：上海文艺出版社发行中心
印刷　上海雅昌艺术印刷有限公司
开本：889×1194　1/32
印张：6　彩插 10
印次：2022 年 12 月第 1 版　2022 年 12 月第 1 次印刷
书号：ISBN 978-7-5535-2671-3/TU.018
定价：78.00 元